The Economics of Refuse Collection

The Economics of Refuse Collection

Peter Kemper
and
John M. Quigley

with the assistance of
Henry Hansmann and Debra Stinson

Ballinger Publishing Company • Cambridge, Mass.
A Subsidiary of J.B. Lippincott Company

This book is printed on recycled paper.

International Standard Book Number: 0-88410-604-7

Library of Congress Catalog Card Number: 75-38941

Printed in the United States of America

Library of Congress Cataloging in Publication Data

Quigley, John M.
The economics of refuse collection.

Bibliography: p.
Includes index.
1. Refuse and refuse disposal—United States. 2. Municipal services—United States. I. Kemper, Peter, 1944, joint author. II. Title.
HD4483.K45 363.6 75-38941
ISBN 0-88410-604-7

Contents

List of Tables

List of Figures

Acknowledgments

A number of individuals contributed in important ways to the research reported in this book. We are especially grateful to William Gamble of the Public Works Department in New Haven, Connecticut; to Bernard Batycki, David Smith and Ray Zavigny of the Public Works Department in Hartford, Connecticut; and to Charles Kurker of the Connecticut State Department of Environmental Protection for making data available to us and for providing other assistance in understanding refuse collection operations. Countless other government officials and private firms generously contributed their time completing questionnaires and answering our questions. We are grateful for their cooperation.

We also appreciate the work done by a number of individuals who acted as research assistants in gathering, verifying and processing data from several diverse sources. We acknowledge the dedicated research efforts of Sara Ford, Johanna Zimmerhart, and Jo Ruggles, and the typing and clerical assistance of Christie Miner and Janice Clancey.

Our debt to Henry Hansmann and Debra Stinson, acknowledged on the title page, is substantial. Henry gathered and analyzed comparative data on municipal budgets with care and precision. Debra assembled much of the published data, collected the questionnaire data, and carefully edited many drafts. Without their patience and persistence this research could never have been completed.

Financial support for this research was provided by the National Science Foundation under grant GI 39589 and by additional funds from Swarthmore College and the Institution for Social and Policy Studies at Yale University. We are grateful to Robert Goldman of

NSF, John Perry Miller, and Charles Lindblom of Yale for encouragement in this research endeavor.

Finally, the manuscript benefited from a careful reading by Eric Hanushek, James Ohls, Richard Murnane, Richard Nelson and Mark Willis. The authors, of course, bear full responsibility for any remaining errors and omissions.

Peter Kemper
John M. Quigley

The Economics of Refuse Collection

Chapter One

Policy Issues in the Provision of Local Services

> [M]ost economists have not turned the spotlight of investigation on the detail of city finances; they have been dissuaded from that venture by the lack of a data base and the need to start by collecting the required information from one public market place or city after another. But progress in policy guidance can only be made if detailed investigations are undertaken. This means fewer reviews of general concepts and far more concentration on the specifics of public products and pricing policies . . . and the optional methods that could be applied to improve the incentive generated by those pricing practices.[1]

Despite the fact that local government expenditures are more than $100 billion annually—more than $500 for every citizen—there is increasing dissatisfaction with the provision of public services, especially in urban areas. Complaints about the level and quality of service provision seem endless, ranging from allegations that our schools fail to educate our youth to charges that mass transport policies are misdirected and that police protection is undersupplied. There is also increasing concern about inequities in the provision of urban services; suits charging that the provision of public services discriminates against lower income households have been brought in more than half the states. Finally, rapid growth in the cost of public services has created dissatisfaction with the slow rate of growth of productivity in the public sector.

In short, public dissatisfaction with service provision involves all three of the most basic criteria of sector performance: (1) efficiency; (2) equity; and (3) progressivity. While public dissatisfaction with the performance of the local government sector is most apparent during recession-induced budget crises, such complaints are not likely to end with economic recovery.

Major Policy Issues

Of the many policy decisions which affect performance in any sector, be it education, fire protection, or refuse collection, one of the most important is the choice of sector organization. For any particular urban service, a municipality may choose to allow the operation of private markets to determine the level and quality of output, or it may choose to supply the service itself through a municipal agency. Nor are these the only options; public authorities may choose to regulate the private market, to franchise the provision of the service, or to contract for service provision with a private firm.

For many goods and services produced in the economy, reliance upon markets and the forces of competition is an adequate way of satisfying consumer demands. If these sectors perform well, it is because prices signal the value of resources to agents in the economy, because competition insures that outputs are produced at minimum cost, and because any socially desired redistribution of income can be accomplished by other government policies independent of that sector.

On the other hand, if private market organization has failed, due to externalities or scale economies, or if notions of ethical distribution are violated in the provision of a local service, public authorities may feel compelled to intervene—to supply the service themselves or to contract for its provision. The decision to intervene is hardly an automatic one in these circumstances, however, because the cure may be worse than the disease. However imperfect provision through the market may be, the alternative forms of government intervention are also likely to be imperfect.

Having chosen to intervene in the provision of a service, decision-makers face several additional choices. They must select the type and level of output and decide on a financing mechanism. Without market prices to signal consumer preferences, the choice of output level is a difficult one, especially when a change in output levels significantly affects the distribution of income. While most public services are financed with property tax revenues and are provided to residents at zero marginal cost, alternative financing methods should also be considered. For some services, zero marginal cost pricing is the only feasible alternative, but for others financing by user charges may be not only possible but also efficient, since the charges would simulate market prices.

In addition, if the local government chooses to produce the service itself, it faces all the operational decisions faced by private firms (except possibly sales promotion). The public agency must choose among alternative technologies, input mixes, and methods of distribution so as to minimize cost. It must find a means of providing management with accurate measures of performance, accurate accounting of costs and—most important—incentives for efficient management and the maintenance of quality. It must define agency responsibilities so that agencies are manageable, but interagency spillovers are minimized. It must find ways to encourage technological innovations which increase productivity. It must provide labor with incentives for productive work and engage in labor relations that avoid exorbitant wage increases and work stoppages.

Informed choices in each of these areas require detailed information. Decision-makers need to know the nature of consumer demand for the particular service as well as the nature of production technology and costs. In addition, they need to assess the past performance of alternative forms of sector organization and to judge its relevance to their particular circumstances. In the absence of a body of research providing evidence, these choices will still have to be made, and they will be decided on the basis of subjective experiences or informed guesses.

To address these issues in a serious way requires a substantial research investment, even for a single local service. Because local services are heterogeneous, because output is difficult to measure, because spillovers and externalities are important, and because reliable data are difficult to obtain, it is not surprising that the body of existing analysis is rather small. Moreover, because the output of the research is itself a public good, it will be under-supplied if financed at the local level.

The focus of this book is on the analysis of a particular local service, residential refuse collection. Unlike most local government services, the output is reasonably well-defined and measurable. Moreover, many types of market organization can be observed in practice. Consequently, the analysis of this particular sector provides an unusually good opportunity to investigate the conceptual issues which are common to virtually all local services provided publicly and privately, and to investigate their importance in the light of empirical evidence.

Beyond methodology that might be generalized to other sectors, there are substantive reasons for the choice of this sector for analysis. Residential refuse collection is a highly visible activity. Local residents, particularly in urban areas, are acutely aware of the level of service provided, and they make any dissatisfaction known. As the ecology and conservation movements have made more Americans conscious of alternatives, there have been numerous proposals for reorganizing the sector, for changing disposal practices, and for making the recycling of solid wastes mandatory. Although this analysis is concerned with the collection of solid wastes rather than their disposal, there are important interdependencies between the two activities. Consequently, improved understanding of refuse collection is needed for informed decisions concerning disposal.

The Scope of the Problem

Perhaps the most significant reason for focusing on this service is that solid waste handling is a large and rapidly growing sector of the economy. It is estimated that $4.5 billion is spent annually in the United States on refuse collection and disposal; of this, about $1.7 billion is spent by public authorities in local civil divisions.[2] Although recent environmental concern has focused attention on waste disposal, it is solid waste collection that currently commands about 80 percent of the public resources devoted to solid waste management, and a substantially higher fraction of the private resources consumed.[3] Crude estimates indicate that, in 1968, residential collection costs averaged $17.60 per ton in those cities and towns providing public collection services. In jurisdictions with populations in excess of 500,000 the average was $24.05.[4]

Table 1–1 presents crude estimates of the amount of refuse (the generic term for all types of solid waste materials) generated annually in the United States from various sources. The estimated total includes the garbage (which refers to organic food wastes) and the rubbish (which refers to organic wastes such as paper and leaves as well as most inorganic materials such as glass and metals) generated from household and commercial sources, as well as the waste products produced by industry and agriculture.

Table 1-1. Estimates of Solid Waste Generation in the United States in 1967.

Source	*Solid Wastes*			
	Pounds per Capita per Day		*Million Tons per Year*	
Local Sector	7.0		256	
Households		3.5		128
Commerce		2.3		84
Municipal, Government		1.2		44
Industrial and Mineral	33.8		1236	
Agricultural	58.0		2115	
Vegetation		15.0		552
Animal		43.0		1563
Federal	1.2		43	
Total	100.0		3650	

Source: Ad Hoc Group for Office of Science and Technology, *Solid Waste Management: A Comprehensive Assessment of Solid Waste Problems, Practices, and Needs* (Washington, D.C.: Office of the President, 1969), Table 7, p. 18.

As the table indicates, the total amount of refuse generated locally from household and commercial sources in the United States is estimated at a staggering 256 million tons per year. Measured by weight, the annual "output" of solid waste is about two times the annual output of steel and five times that of coal. In addition to this waste generated by households and commerce, it is estimated that another 3.4 billion tons of waste are produced by industry and agriculture.

These estimates of refuse generated are quite crude. The 1968 national survey, from which these data are derived, contains information on the solid wastes generated in communities containing only about half of the U.S. population.[5] Of the 6,000 communities which participated in the survey, less than half were able to supply estimates of the tonnage of refuse generated or collected.[6] Nevertheless, excluding federal, industrial, and agricultural waste products, the best estimate of the amount of refuse generated in the United States is 7 pounds per person per day.

If solid wastes were not costly to collect and expensive to dispose of, the consequences of this enormous output would be of little public interest. But other estimates indicate that, even though the overwhelming proportion of refuse generated by industry and agriculture is transported and disposed of by the producer, most of the refuse generated by the local sector is collected and disposed of by local government, either through municipal agencies themselves or by contract. For example, of the 7 pounds per person per day generated, it is estimated that 5.3 pounds are actually collected by firms and local government agencies, including 3 of the 3.5 pounds generated by households. The average local collection of 5.3 pounds per person per day includes an estimated collection of 3.9 pounds in rural areas and 5.7 pounds in urban areas.[7]

Practically nothing is known about the trends in the amount of refuse generated in the local sector. The American Public Works Association reports uncertain estimates of a per capita daily refuse generation of about 2.7 pounds in 1950.[8] If these figures are at all accurate, they suggest that per capita refuse generation more than doubled between 1950 and 1967. Some reliable estimates for particular cities tend to confirm these trends. For example, per capita refuse generation in New York City increased by 78 percent during the decade 1955-1965; in Los Angeles per capita generation increased by 51 percent during the decade 1958-1968.[9]

If there is uncertainty about the rate of growth in the amount of refuse generated, there is little doubt that the amount of resources devoted to solid waste collection has increased dramatically. As an indication of the growth of the sector, the U.S. Census of Population reported that there were about 106,000 employees in the sanitary services occupations in 1950. In 1960 the number had grown to 147,000 employees. By 1970 there were more than 178,000 persons employed in the sector, a growth of 68 percent during a twenty-year period.

Accompanying the substantial growth in the quantities of solid wastes generated and in the resources devoted to its removal, has been a change in the composition of refuse. Table 1–2 suggests that the proportion of solid waste that is paper has increased and the proportion of garbage and ashes has declined.

These secular changes in both the quantity and composition of solid waste probably result from continuing urbanization, changes in household incomes and tastes, and technological changes. As populations have shifted from rural to urban areas, the demand for collec-

Table 1-2. Composition of Municipal Refuse for Selected Cities over Time.

	Organic			*Inorganic*		
	Garbage	*Paper*	*Total Organic*	*Metal*	*Ashes*	*Total Inorganic*
1939 - New York City[a]	17.0%	21.9%	44.7%	6.8%	43.0%	55.3%
1956 - Chicago[b]	5.1	57.6	67.3	14.4	15.5	32.7
1957 - Chicago[b]	3.8	51.0	66.4	9.9	19.4	33.7
1958 - Chicago[b]	2.4	60.5	69.8	8.3	15.6	30.2
1967 - Berkeley[c]	25.1[d]	44.6	69.7	8.7	-	30.3
1970 - Jefferson Co., Ky.[c]	20.5[d]	56.4	76.9	10.6	1.1	23.1
1972 - Cincinnati[e]	23.9[d]	48.2	72.1	9.0	5.7	27.9

[a]Casimir A. Rogus, "Refuse Quantities and Characteristics," *American Public Works Association Proceedings* (National Conference on Solid Waste Research, December, 1963), pp. 17–27.

[b]American Public Works Association, *Municipal Refuse Disposal* (Chicago: Public Administration Services, 1961).

[c]American Public Works Association, *Municipal Refuse Disposal* (Chicago: Public Administration Services, 1970), p. 49.

[d]Including all nonpaper organic matter.

[e]George R. Davidson, Jr., *A Study of Residential Solid Waste Generated in Low-Income Areas*, Report SW 83ts (Cincinnati: U.S. Environmental Protection Agency, 1972), p. 9.

Source: American Public Works Association, *Refuse Collection Practices* (Chicago: Public Administration Services, 1966), p. 38.

tion has increased, if only because households are less able to dispose of refuse by burning or composting on their own properties. In addition, rising household incomes have led to increased consumption of all types of goods, especially disposable goods, which has increased the quantity of refuse generated as a by-product of consumption.

As the economy has become more interdependent and packaging more mechanized, the amount of packaging per unit output has increased. There has also been increased reliance on prepared and frozen foods, resulting in a decreased garbage content of residential refuse but an increased amount of paper and cardboard packaging. Other aspects of technological change have altered the composition of household refuse. Use of garbage disposers has further reduced the raw garbage content of residential refuse. Reliance on gas and oil for home heating, as opposed to wood and coal, together with restrictions on open burning in urban areas, has reduced the ash content of residential refuse. Finally, simple increases in population have resulted in substantial additions to the total volume of refuse generated and collected.

In short, the size of the sector, its continued growth, and increased concern for the environment make refuse collection a significant policy problem that is of interest in its own right. The remainder of this book is devoted to a detailed investigation of residential refuse collection. In addition to providing information to those concerned with residential refuse collection policy, it is hoped that the analysis provides some insights for policy and research in other local government services.

PLAN OF ANALYSIS

The fundamental concerns underlying the analysis are matters of sector organization and financing: the implications of alternative forms of organization (provision by a government agency, by a private organization under government contract, or by a free or regulated private market), the merits of user charges as compared to property tax financing, and the institutional incentives for the choice of financing.

While these questions motivate the research, they cannot be answered in the abstract. Documentation of existing practices and a careful analysis of the basic cost and demand conditions for refuse collection form a necessary foundation for intelligent discussion of these issues. Besides relying upon survey information for the nation as a whole, the results reported in this volume draw upon several surveys of refuse collection operations in 1972 in the state of Connecticut. Using differing methodologies, data were obtained from each Connecticut jurisdiction providing municipal collection service, from each jurisdiction providing service by contract with a private firm, and from a sample of private firms providing refuse collection services in other jurisdictions. In addition, some of the empirical analysis makes use of internal records on collection operations maintained by the public works departments in two large cities, New Haven and Hartford, Connecticut.

Chapter 2 contains an analysis of the structure of costs, which provides the basis for the extended discussion of sector organization in Chapter 3. Chapter 4 presents an empirical analysis of the demand for refuse collection services, which, together with the cost analysis in Chapter 2, provides the foundation for the discussion of financing methods in Chapter 5.

Thus, by analyzing cost and demand, Chapters 2 and 4 provide a somewhat technical foundation for the applications in Chapters 3 and 5. Readers primarily interested in applications may wish to skim over Chapters 2 and 4. Those interested in improving upon the research design and the methods of analysis may wish to read the appendices, which document the research in some detail.

The final chapter presents a summary of the empirical findings, based largely on the analysis of data from the state of Connecticut, and suggests some implications for policy and further research.

NOTES

1. Selma J. Mushkin, ed., *Public Prices for Public Products* (Washington, D.C.: The Urban Institute, 1972), p. 426.

2. R. J. Black, et al., "The National Solid Wastes Survey: An Interim Report," Paper presented at the Annual Meeting of the Institute for Solid Wastes, American Public Works Association, October 24, 1968, pp. 13-15.

3. Ralph Stone and Company, Inc., Engineers, *A Study of Solid Waste Collection Systems Comparing One-Man with Multi-Man Crews*, U.S. Public Health Service, Publication #1892 (Los Angeles: U.S. Department of Health, Education and Welfare, 1969).

4. R. J. Black, et al., op. cit.

5. Anton J. Muhich, Albert J. Klee, and Paul W. Britton, *Preliminary Data*

Analysis: 1968 National Survey of Community Solid Waste Practices, Public Health Service Publication #1867 (Cincinnati: U.S. Department of Health, Education and Welfare, 1968).

6. Ibid., p. 12. Furthermore, for each item contained in the 1968 survey, the participating communities were asked to evaluate the quality of the data they supplied. Some 56 percent of the respondents indicated that their estimates of the amount of refuse generated were of "poor" quality and only 6 percent of the respondents indicated that their estimates were of "good" quality.

7. Anton J. Muhich, "Sample Representativeness and Community Data," in *The National Solid Waste Survey: An Interim Report*, by Ralph J. Black, Anton J. Muhich, Albert J. Klee, H. Lanier Hickman, Jr., and Richard D. Vaughan (Cincinnati: U.S. Department of Health, Education and Welfare, 1968).

8. American Public Works Association, *Refuse Collection Practices* (Chicago: Public Administration Services, 1966), p. 25.

9. American Public Works Association, Institute for Solid Wastes, *Municipal Refuse Disposal* (Chicago: Public Administration Service, 1970), p. 27.

Chapter Two

The Cost of Refuse Collection

Almost every decision faced by local officials concerned with refuse collection and disposal affects costs. In order to make rational choices, officials need quantitative estimates of the costs of alternative policies. For example, the fundamental decision about sector organization requires some understanding those factors affecting the cost of refuse collection and their relative magnitudes. Other basic decisions, such as the choice of type of service and frequency of collection, require informed estimates of the costs of alternatives. In addition, if the government seeks to evaluate alternative methods of financing collection, such as user charges or special service fees, reliable cost data are essential. Estimates of collection costs are also needed at a somewhat more detailed level for operational decisions such as route design and the choice of equipment and crew size. Finally, cost estimates are required to evaluate proposed changes in disposal practices. Moving the disposal site farther from the center of town will increase hauling distances and hence collection costs; proposals for recycling requiring the separate collection of bottles, metal, and other refuse will typically increase collection costs.

In short, informed choices among major policy alternatives and routine operational decisions require knowledge of the relationship between collection costs, service and frequency levels, method of organization and the spatial characteristics of the city.

In this chapter we develop the empirical evidence that might form the basis for some of these decisions. After reviewing the existing evidence, two types of cost functions are estimated. The first analyzes the variations in average collection cost across locations within a single city; the second analyzes variations in average collection cost across cities.

EXISTING EVIDENCE ON COST VARIATIONS

Despite the obvious policy importance of cost relationships for refuse collection, there is little descriptive information suggesting how costs vary with such factors as service level, density, or other factors. Even less information is available about the functional relationship between service outputs and resource costs.

Table 2-1 presents fragmentary budgetary data gathered by Ralph Stone and Company indicating average budget expenses per ton of refuse for three categories of city size. Although the figures are extremely crude and are based on only a few cities, the comparisons suggest that unit costs are higher for larger cities and that back-of-the-house pickup is substantially more expensive than curbside service.

An American Public Works Association (APWA) monograph, *Refuse Collection Practices*, suggests that jurisdictions with high reported costs are often those cities which, "to provide administratively valuable data include all elements of cost in their reported figures."[1] The APWA monograph concludes that in most cases "comparisons of such municipal cost data are worse than useless." A National Academy of Sciences report of 1966 concludes more generally that "a major shortcoming in solid waste management is

Table 2-1. Refuse Collection Costs per Ton for Selected Cities.

		Curbside pickup		*Backyard pickup*	
Population (000's)	*Average collection costs per ton (166 cities)*	*Average cost per ton*	*Sample size*	*Average cost per ton*	*Sample size*
10-100	$ 9.50	$ 8.61	21	$10.71	5
100-500	10.20	8.92	6	15.78	4
500+	24.05	20.71	2	14.09	1
All cities	17.66	9.52	29	13.08	10

Source: Ralph Stone and Company, Inc., Engineers, *A Study of Solid Waste Collection Systems Comparing One-Man with Multi-Man Crews* (Los Angeles; U.S. Department of Health, Education and Welfare, 1969), p. 61.

the lack of accurate quantity and meaningful cost data, making it virtually impossible to evaluate alternative methods, thus perpetuating archaic practices."[2]

A review of the economic literature reveals that there has been only one serious attempt at estimating cost functions for refuse collection, that by Werner Hirsch.[3] Hirsch attempts to estimate a cost function using 24 cities in St. Louis County as units of observation. Hirsch regresses "average annual residential refuse collection and disposal cost per pickup" upon seven measures of scale, service level and financing arrangements. The independent variables for each city are number of pickup units, number of units squared, pickup density (number of pickups per square mile), frequency of collection, and three dummy variables for curbside pickup, private collection, and user-charge financing. Of these variables, only frequency of collection, curbside pickup and user-charge financing significantly affect average cost. Hirsch's analysis implies that there are no economies of scale in refuse collection and disposal activities and, more surprisingly, that there are no economies of density (i.e., that average cost is insensitive to the density of pickups). Elsewhere Hirsch relies upon this evidence to classify refuse collection activity as a horizontally integrated service with a horizontal average cost curve.[4]

There are, however, several reasons for questioning the conclusions of Hirsch's analysis. In the first place, as Hirsch recognizes, the relevant definition of output is tonnage or volume collected, not simply the number of pickups or units served. Only if the average amount of refuse generated per household were constant would the two measures coincide. Second, the combined analysis of collection and disposal ignores an important conceptual distinction between economies of scale and economies of density. Economies of scale would exist if average costs declined with the quantity of output; economies of density would exist if average costs declined with the density of collection over space. One expects to observe economies of scale for refuse disposal activities but no economies of density. For refuse collection, on the other hand, we expect to observe economies of density, but there is a little *a priori* reason to expect economies of scale. By combining the two activities in a single analysis, any economies of scale in disposal and economies of density in collection are likely to be blurred. Finally, budgetary cost data are unlikely to represent the true economic cost of municipally provided service; in particular, the comparisons of costs per unit across cities or between cities with contracted service and those with municipally provided service are subject to substantial measurement error.

In short, a review of existing literature reveals little useful empirical research on the cost of refuse collection. The remainder of this chapter seeks to provide some of that information.

INTRACITY COST FUNCTIONS

We begin by investigating the way costs vary across routes within a single city. Limiting the analysis to one city has the methodological advantage that many factors which affect collection cost are held constant so that a relatively pure test of the effect of other factors can be made. For example, factor prices and the type of service provided are constant within a city, but the characteristics of routes vary within that same city. Thus the intracity analysis provides a good opportunity to measure the effect of route characteristics on collection costs. After developing a conceptual framework, cost functions are estimated using data from two cities, Hartford and New Haven, Connecticut.

Conceptual Framework

To develop a cost function that can be estimated empirically, it is necessary to define the output produced, to examine the production process, and based on that examination, to postulate the theoretical relationship between average cost and those variables which affect it.

Defining the output of solid waste collection activity is easier than for most services provided by municipalities, but it is still not without some complications. The obvious output is the collection of a given amount of refuse at a particular location and its transportation to a disposal site, but there are also other explicit objectives, such as the elimination of health hazards and the reduction of street and yard litter. Presumably the achievement of these objectives is highly (though not perfectly) correlated with the process of removing a given amount of refuse from individual residences. For this reason and because of the difficulty of measuring the other objectives, we use the tonnage collected and transported to the disposal site as our measure of output.

Residential solid waste collection, as presently performed, is a rather simple production process. The inputs typically consist of a compactor truck, a driver, and two laborers. The truck stops, the laborers get the containers and empty them into the compactor, and the truck moves on to the next house. When the truck is filled, the load is driven to the disposal site, the load is dumped, and the truck returns to resume pickups. The total time spent collecting consists of two distinct parts, the "pickup time," which is the time spent emptying cans and driving between houses, and the "haul time," which is the time spent hauling the load to the disposal site, dumping it, and returning to the collection route.

Two things are clear from this description of the production process. First, refuse collection is a labor-intensive process so that, at a given wage rate, the cost per ton will depend in large part on the time it takes to collect each ton. Second, the time it takes to collect each ton depends on the type of service provided and the characteristics of the route, such as density and hauling mileage.

A simple model,[a] based on these two observations, begins by recognizing that average cost per ton can be written as the product of cost per unit time and collection time per ton, and that cost per unit time can be divided into labor and capital cost:

$$AC = \frac{TC}{T} \bullet \frac{T}{Q} = (c_L + c_K)\frac{T}{Q} \qquad (2\text{-}1)$$

where:

AC = average cost
TC = total cost
T = total time spent collecting
Q = total tonnage collected
c_L = cost of labor per unit time, including wages, fringe benefits and direct supervisory costs
c_K = rental cost of trucks per unit time they are actually utilized including interest, depreciation, operation and maintenance

The time per ton can be further divided into the pickup time per ton and the haul time per ton:

$$\frac{T}{Q} = \frac{T_p}{Q} + \frac{T_h}{N} \bullet \frac{N}{Q} \qquad (2\text{-}2)$$

[a]The model developed here is based in part on that in American Public Works Association, Committee on Solid Wastes, *Refuse Collection Practices* (3rd ed., Chicago Public Administration Service, 1966).

where:

T_p = total pickup time
T_h = total haul time
N = total number of trips to the disposal site

Combining (2–1) and (2–2) gives

$$AC = (c_L + c_K)\left(\frac{T}{Q}\right) = (c_L + c_K)\left(\frac{T_p}{Q} + \frac{T_h}{N} \bullet \frac{N}{Q}\right) \tag{2-3}$$

As it stands, equation (2–3) is simply an identity; it becomes a useful behavioral equation when the variables which affect labor and capital costs per unit time, pickup time per ton, and haul time per run are specified. Since the analysis here is limited to one city, we can assume that labor and capital costs are constant across routes. Pickup time and haul time, however, will vary across routes.

A number of factors can be expected to cause variation in pickup time. Of major importance is the residential density of the dwelling units served. The greater the density, the smaller the total mileage that must be traveled to collect a given amount of refuse. Thus, at higher densities, the average cost of refuse pickup should be lower. In addition, size of the crew and the type of containers used will affect the time required to pick up a given quantity of refuse. Finally, weather conditions, especially snowfall, will affect the rate at which crews can work. Formally,

$$\frac{T_p}{Q} = f\,(D, C, CN, S) \tag{2-4}$$

where:

D = density of collection: tons of refuse per pickup mile
C = crew size
CN = type of container used
S = amount of snow on the ground

The time it takes to haul and dump each load depends primarily on the distance from the route to the disposal site. In addition bad weather may affect driving conditions, slowing the speed at which trucks can travel. Formally,

$$\frac{T_h}{N} = g(H, R) \tag{2-5}$$

where:

H = haul mileage
R = rain or snow

Obtaining empirical estimates of equations (2–4) and (2–5) is not a simple task. Conceivably they could be estimated from a time-and-motion study. Observers could accompany each crew, counting the number of cans, clocking the time spent at each task, measuring mileages, and so on. From data collected in this manner, some specification of equation (2–3) could be approximated.

There are several serious problems with this approach, however. In the first place, such a study would be extraordinarily expensive even to obtain estimates for a single city. Second, workers are notoriously resistant to such observations; unionized workers in the public sector are unlikely to be less resistant. Third, the effect of being observed may change behavior, making direct estimates unreliable.

An alternative, which avoids the problems of a time-and-motion study, is the statistical estimation of the model using multiple regression analysis. This approach is used to estimate cost functions for Hartford and New Haven.

Hartford Cost Function

The city of Hartford provides once-a-week collection and "roll-out" service (i.e., cans are rolled out from behind the house to the curb by a special crew) to all city residents.[b] Collection activities are organized so that, on any day, three routes are served, giving a total of fifteen distinct routes. Each route is served by five trucks, each with a crew of two loaders and a driver. In addition, roll-out laborers, who work independently, average two per truck. If the

[b]In addition, two special services are provided: daily collection in the central business district and weekly "bulky" collection to residences for items such as furniture and appliances too large for collection by the regular compactor trucks.

roll-out service is viewed as independent of collection, the operation uses the technology typical of most towns in Connecticut and elsewhere.[c]

From time cards kept by the Public Works Department, it was possible to calculate, for each day and route, the aggregate amount of time spent collecting refuse, the time spent hauling refuse, and the total amount of refuse collected on each route.[d] From maps supplied by the Public Works Department, it was possible to estimate the mileage along each route and its distance to the disposal site. In addition, for each day of the year, weather conditions (snow and rain) were obtained from the U.S. Weather Bureau. Together these sources provided a sample of 519 observations, one for each "route-day" (i.e., for each route served on each day) during 1972, which we have used to estimate equations (2–4) and (2–5).[e]

Table 2–2 presents the estimated relation between pickup time per ton and density, crew size, and the amount of snowfall.[f] The equations differ in their specifications of the relationship beween pickup time and density. In particular, the equations are estimated for a quadratic and a reciprocal relationship as well as for the simple linear form. For the regression results shown, the individual coefficients are highly significant statistically, and the equations as a whole explain between 31 and 42 percent of the variation in pickup times. A six-inch snowstorm increases the required time by about 25 percent. For this sample, the average pickup time per ton is about 21 minutes. An additional worker assigned to the roll-out crew reduces average time by about 6 percent.

Most important, the regressions provide strong evidence of economies of density in residential refuse collection. The elasticity estimates shown in Table 2–2 indicate that a 1 percent increase in

[c]Even including roll-out operations, the production process appears to use inputs in fixed proportions; however, there is some variation in the number of roll-out laborers along a route due to absences, and if enough men are absent, the number of trucks is reduced from five to four.

[d]The source and construction of these variables is described in detail in Appendix A.

[e]The means and standard deviations of all variables used and frequency distributions for the dependent variables appear in Appendix D.

[f]Information on the type of container used was not available. Omission of this variable may have either of two effects. On the one hand, use of plastic bags, which speeds pickup, is probably negatively correlated with density, which would bias the results against observing economies of density. On the other hand, use of special five-yard containers, which also speeds collection, is probably positively correlated with density, which would bias the results in the opposite direction.

Table 2-2. Pickup Time Regressions for Hartford (dependent variable: pickup time per ton; number of observations: 519)[a]

Density (tons per pickup mile)	−0.76 (11.8)	−3.07 (10.1)	—
Density squared	—	0.15 (8.1)	—
Reciprocal of density	—	—	39.50 (15.9)
Crew size[b]	−0.28 (3.8)	−0.25 (3.5)	−0.25 (3.7)
Snow (inches at 7:00 a.m.)	0.98 (7.2)	0.91 (7.1)	0.91 (7.3)
Constant	32.88 (17.7)	40.22 (20.4)	20.52 (11.4)
R^2	0.31	0.39	0.42
Elasticity of pickup time with respect to density (evaluated at means)	−0.24	−0.36	−0.28

[a]t-ratios in parentheses.

[b]Along each route, including driver, loaders, and roll-out crew members.

collection density reduces average pickup time by roughly 0.3 percent. The regression results indicate, moreover, that the magnitude of these economies of density declines as collection density is increased. For example, the reciprocal specification, which explains 8 percent more of the variance in collection times than the simple linear representation, suggests that as collection density is increased from two to three tons per pickup mile, average pickup time is reduced from 40.8 to 34.3 minutes per ton, or by 16 percent. However, as collection density is increased from seven to eight tons per pickup mile, average pickup time is reduced from 26.7 minutes to 26.0 minutes per ton, a change of only 3 percent.

These findings are illustrated in Figure 2–1, which presents a plot of the pickup time per ton of refuse versus collection density and the estimated regression equation, after controlling for the effect of snowfall and crew size. Refuse collection is clearly subject to important economies of density.

Table 2–3 presents the regression of haul time per run on haul mileage and weather conditions. Since for any route the hauling distance is constant for all days, the model was estimated from observations on the average haul times for each of the 15 routes for

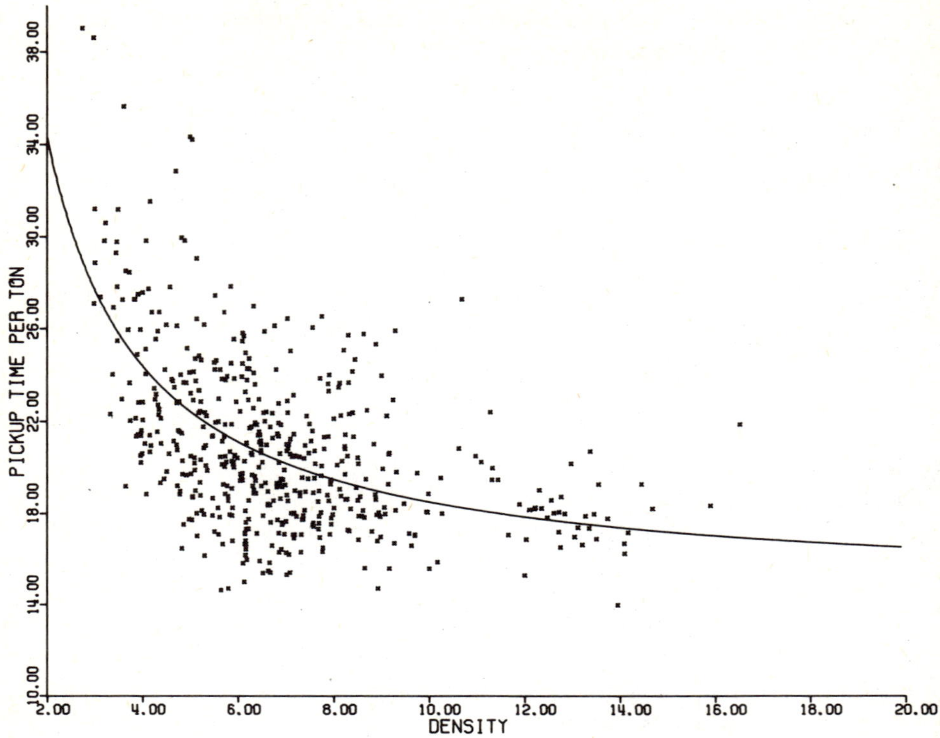

Figure 2–1. Average Pickup Time Versus Density for Hartford Sample

two classes of weather conditions (days with precipitation and days without precipitation). Table 2–3 also reports separately the results for the 15 routes on days without precipitation.

The results indicate that the mileage traveled explains some 20 to 37 percent of the variation in average hauling time and is statistically significant. The magnitude of the effect of haul mileage on haul time is unreasonably small; an increase of 1 mile increases haul time by only 0.33 minutes. If the additional mile were literally traversed in 0.33 minutes, this would imply a haul speed of 180 miles per hour! A possible explanation of this unreasonably large esimate is that "break time" is not independent of "haul time"; if a crew has a long haul it may take a shorter break, causing a negative correlation between break time and haul mileage and a downward bias on the coefficient of haul mileage.

The results further suggest that precipitation slightly increases the time required to haul and dump refuse. When total haul mileage is

Table 2-3. Haul Time Regressions for Hartford (dependent variable: haul time, in minutes, per round trip).[a]

	All weather conditions[b] (Number of observations: 30)				*No precipitation (Number of observations: 15)*	
Haul mileage						
Total haul miles	0.39 (5.2)	0.39 (5.3)	—	—	0.33 (4.7)	—
Along interstates	—	—	0.39 (4.7)	0.39 (4.8)	—	0.23 (3.4)
Along city streets	—	—	0.41 (3.5)	0.41 (3.5)	—	0.65 (6.4)
Rain (1 if rain, 0 otherwise)	—	0.56 (1.6)	—	0.56 (1.6)	—	—
Constant	31.13 (45.6)	30.94 (45.0)	31.09 (43.4)	30.90 (42.8)	31.48 (49.4)	30.70 (50.3)
R^2	0.20	0.21	0.20	0.21	0.23	0.37

[a] *t*-ratios in parentheses.

[b] There is one observation on the average haul time for each of the 15 different routes for days with and without precipitation.

Table 2-4. Estimates of Refuse Collection Costs per Minute for Hartford in 1972.[a]

A. Overhead cost not allocated to labor and capital:		
Category	*Total Cost*	*Cost per minute*
Overhead	$ 29,000	$ 0.02
Direct labor[b]	1,260,000	.79
Direct capital	383,000	.24
Total	$1,672,000	$ 1.05
B. Overhead costs allocated proportionally to labor and capital costs:		
Category	*Total Cost*	*Cost per minute*
Labor	$1,282,000	$ 0.80
Capital	390,000	0.25
Total	$1,672,000	$ 1.05

[a]See Appendix B for methodology and assumptions.

[b]Of the total labor cost, approximately $340,000 is accounted for by the roll-out crews.

broken down into mileage on interstate highways and mileage on city streets, surprisingly little difference between the two coefficients is noted. However, when the model is estimated separately for days without precipitation, the results suggest that not only does rain decrease hauling speed, but it also causes a greater decrease in hauling speeds along interstate highways.

The overall explanatory power of the statistical models in Table 2–3 is, in any case, quite modest. Besides reflecting random variations in the length of crew breaks, this undoubtedly reflects the fact that hauling times are measured inexactly by drivers; quite often times are rounded to the nearest five minutes, creating substantial measurement error.

To use the empirical estimates in Tables 2–2 and 2–3 to obtain the cost function for residential refuse collection in Hartford, estimates of labor and capital costs per unit time are required. A detailed study of Hartford's expenditures was required to estimate the public resources devoted to refuse collection. This analysis, which is described in detail in Appendix B, provided estimates of annual labor, capital, and overhead costs for regular collection. As summarized in Table 2–4, the total cost of refuse collection in Hartford is estimated to be $1,672,000, or about $1.05 per minute. Of this, about three fourths consists of direct labor expenses. An additional

$383,000 or $.24 per minute consists of direct capital expenses. In addition, a small sum of $29,000 or $.02 per minute consists of overhead expenses. Labor and capital costs are estimated to be $.80 and $.25 per minute respectively, when the small overhead expenditure is allocated in proportion to direct labor and capital expenditures.[g]

Using these estimates together with the average number of trips to the incinerator per ton of refuse, .154, and the regression results in Tables 2–2 and 2–3, the parameters of the cost function for Hartford can be expressed as:

$$AC = (c_L + c_K)\left(\frac{T_p}{Q} + \frac{T_h}{N} \bullet \frac{N}{Q}\right) \tag{2-6}$$

$$= 21.44 + 41.28\left(\frac{1}{D}\right) - 0.26C + 0.95S + 32.08\left(\frac{N}{Q}\right)$$

$$+ 0.68M_s\left(\frac{N}{Q}\right) + 0.24\,(M_I)\left(\frac{N}{Q}\right)$$

$$= 26.38 + 41.28\left(\frac{1}{D}\right) - 0.26C + 0.95S + 0.10M_S + 0.04M_I$$

where:

AC = average cost per ton of refuse
D = density of residential collection (tons per pickup mile)
S = snow on ground at 7:00 a.m. (inches)
M_s = haul miles on city streets from route to incinerator
M_I = haul miles on interstate highways from route to incinerator
C = crew size
$\frac{N}{Q}$ = number of trips to incinerator per ton of refuse

Discussion of these results is postponed until a later section.

New Haven Cost Function

It was possible to conduct an independent estimate of a cost function using data from New Haven. The operation of refuse collection in New Haven is similar to that in Hartford, although the type of

[g]The labor costs are much higher than the actual wage rates paid laborers because they include indirect labor costs, such as fringe benefits and supervisor wages, and because they are based on time actually worked, not including the paid time spent before collection begins or after the last load is dumped.

service differs. The city of New Haven provides twice-a-week, curbside collection to all residential structures containing five or fewer dwelling units. The inputs consist of a driver, two laborers and a compactor truck. In contrast with the organization of refuse collection in Hartford, each truck and crew operates independently, serving one of 11 routes each day. Twice-a-week service is provided in six work days; thus there are 33 distinct collection routes, each of which is served by a single truck and crew.

Although the collection operation is, if anything, simpler in New Haven than in Hartford, the records on collection times maintained by the Public Works Department are not as detailed. Nonetheless, using the available records, a route map, and weather data from the U.S. Weather Bureau, a sample less detailed but similar to that developed for Hartford was gathered for the fiscal year 1972. In comparison with the data analyzed for Hartford, haul times and pickup times are not indicated separately, and an additional assumption required to compute the total collection time introduces some measurement error in that figure. The procedures used to construct the sample are described in Appendix A.

These data do not permit the estimation of an aggregate cost function identical to that estimated for Hartford because they do not permit the disaggregation of total time into pickup time and haul time. Instead, the model applied to the New Haven sample combines the separate pickup and haul time equations (2–4) and (2–5) into a single total time equation:

$$\frac{T}{Q} = \frac{T_p}{Q} + \left(\frac{T_h}{N}\right)\left(\frac{N}{Q}\right) = f(D, C, S) + \left(\frac{N}{Q}\right) g\,(H, R) \qquad (2\text{–}7)$$

where the variables are defined as before.

For the New Haven operation, however, roll-out service is not provided, thus crew sizes (C) are constant. In addition, since the rainfall measure (R) did not appear to be significant in the Hartford regressions, we ignore it in the New Haven analysis. Because the equations reported in Tables 2–2 and 2–3 are linear in their parameters, the analogous model for New Haven becomes

$$\frac{T}{Q} = \alpha_1 + \alpha_2 D + \alpha_3 S + \alpha_4 \left(\frac{N}{Q}\right) + \alpha_5 \left(\frac{NH}{Q}\right) \qquad (2\text{-}8)$$

where the α's are the parameters to be estimated.

As before, density is included in the regressions in several alternative ways: linear, quadratic, and reciprocal. As in the Hartford analysis, haul miles are disaggregated into miles along interstate highways and miles along city streets.[h]

Table 2–5 summarizes the regression results and indicates that the simple model of refuse collection explains between 45 and 55 percent of the variance in average collection times. With the exception of the variable representing snow on the ground during collection, all the coefficients are highly significant. In contrast to the Hartford results, haul mileage significantly affects the total time required to collect a ton of refuse. Depending on the specification, an additional mile of hauling per ton increases collection time by 3 to 5 minutes (per ton). When haul mileage is broken down into mileage on city streets and interstate highways, the results again suggest that average times are reduced if more hauling is done on interstate highways. In contrast to the analysis for Hartford, the estimated coefficients imply hauling speeds that are more reasonable, 10 mph on city streets and 28 mph on interstate highways.

The results for New Haven further substantiate the importance of economies of density in residential refuse collection. The elasticity calculations shown in Table 2–5 indicate that a 1 percent increase in collection densities reduces average collection time by from 0.2 to 0.4 percent, depending on the specification. Again, the magnitude of these economies of density declines as collection density is increased. As in the Hartford regressions, the best fit to the data is obtained by the reciprocal function, indicating a hyperbolic relationship between average collection time and collection density. These results are illustrated in Figure 2–2, which presents a plot of the average collection time versus collection density together with the estimated regression equation.[i]

To use these empirical results to obtain the cost function for residential refuse collection in New Haven, estimates of capital and labor costs per unit time are again required. A study of New Haven's expenditures, similar to but less detailed than that for Hartford, provided estimates of the economic costs of labor, capital, and overhead for New Haven.[j] As shown in Table 2–6, the total annual cost of residential refuse collection is estimated to be about $641,000 or

[h]Means, standard deviations, and frequency distributions of the variables used are reported in Appendix D.

[i]Again, the plotted values are adjusted for the effect of the other explanatory variables in regression.

[j]This study is described in Appendix B.

Table 2-5. Total time Regressions for New Haven (dependent variable: total collection time per ton; number of observations: observations: 2791).[a]

Density:						
Density (tons per pickup mile)	−2.392 (24.15)	−2.248 (22.32)	−6.636 (19.18)	−6.481 (12.83)		
Density squared			0.694 (12.77)	0.692 (12.83)		
Reciprocal of density					14.870 (34.90)	14.370 (33.22)
Snow (inches at 7:00 a.m.)	0.144 (0.95)	0.116 (0.77)	0.100 (0.60)	0.072 (0.49)	0.047 (0.34)	0.027 (0.20)
Runs per ton	47.580 (16.18)	44.170 (14.90)	45.270 (15.81)	41.910 (14.51)	37.870 (13.89)	35.310 (12.85)
Haul miles:						
Total haul miles per ton	5.073 (12.01)		3.809 (9.02)		3.284 (8.31)	
Along city streets		8.286 (12.77)		6.990 (10.95)		5.924 (9.82)
Along interstates		3.627 (7.64)		2.382 (5.05)		2.132 (4.84)
Constant	16.400 (32.39)	16.250 (32.31)	22.870 (32.38)	22.710 (32.37)	5.800 (15.98)	6.158 (16.82)
R^2	0.447	0.453	0.475	0.483	0.533	0.538
Elasticity of total time with respect to density (evaluated at mean)	0.242	0.227	0.361	0.346	0.308	0.298

[a] t-ratios in parentheses.

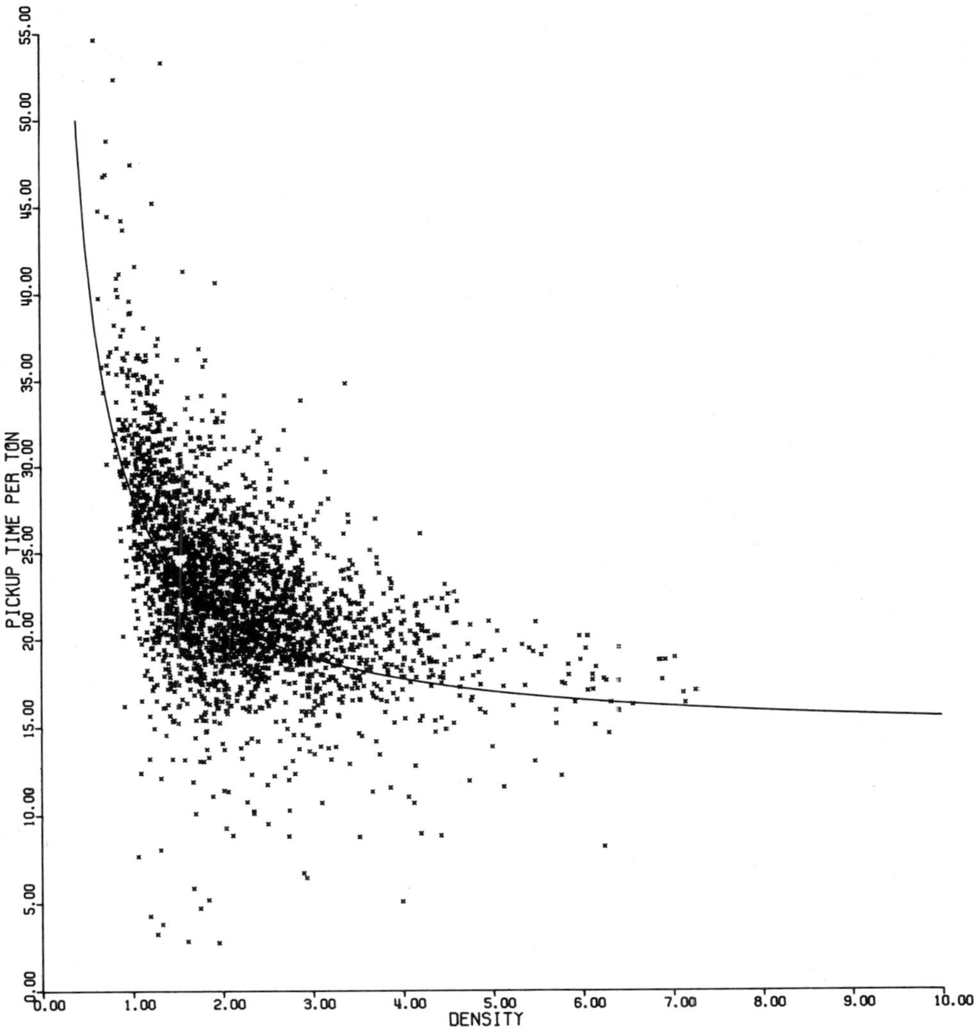

Figure 2-2. Average Collection Time Versus Density for New Haven Sample

roughly $.83 per minute.[k] As compared with Hartford, collection costs are lower in New Haven, if only because roll-out service is not provided. When the small overhead expenditure is allocated in proportion to direct labor and capital expenses, it is estimated that

[k]As noted above, these labor costs should not be compared with hourly wages because they include indirect labor costs and are based on actual collection time.

Table 2-6. Estimates of Refuse Collection Costs per Minute for New Haven in 1972.[a]

A. Overhead cost not allocated to labor and capital:

Category	*Total Cost*	*Cost per minute*
Overhead	$ 19,000	$ 0.02
Direct labor	477,000	.62
Direct capital	145,000	.19
Total	$641,000	$ 0.83

B. Overhead costs allocated proportionally to labor and capital:

Category	*Total Cost*	*Cost per minute*
Labor	$491,000	$ 0.64
Capital	150,000	.19
Total	$641,000	$ 0.83

[a]See Appendix B for assumptions and methodology.

the cost of labor used is about $.64 per minute, and the cost of capital is about $.19 per minute.

Using these estimates, together with the average number of trips to the incinerator per ton of refuse, .159, the parameters of the cost function for New Haven, based on the regression results in Table 2–5, may be expressed as:

$$AC = (c_L + c_K)\left(\frac{T}{Q}\right) \tag{2-9}$$

$$= 5.11 + 11.92\left(\frac{1}{D}\right) + 0.02S + 29.31\left(\frac{N}{Q}\right) + 4.91\left(\frac{N}{Q}\right)M_S + 1.77\left(\frac{N}{Q}\right)M_I$$

$$= 9.77 + 11.92\left(\frac{1}{D}\right) + 0.02S + 0.78M_S + 0.28M_I$$

Comparison of Intracity Results

The results of the separate analyses of collection operations in Hartford and New Haven are surprisingly consistent with prior expectations and with one another, despite the differences in the services performed.[1]

[1]Hartford, the larger city, provides daily collection service to central business district establishments, while New Haven provides no collection service to commercial establishments. Hartford provides weekly service to all dwelling units in the city; New Haven provides twice-a-week service to all single-family homes and small multifamily structures, but provides no service to apartments

The average cost of refuse collection varies with the haul mileage between the collection route and the incinerator, although the magnitude of the effect is substantially greater in New Haven than Hartford. In addition, if more of the hauling mileage takes place on city streets than on interstate highways, average collection costs are increased. In Hartford, where roll-out service is provided, snow on the ground increases the average collection time; in New Haven, where curbside service is provided, the results indicate that snow has no discernible effect on collection time.

For both cities, there are marked economies of density in refuse collection; that is, average cost decreases as collection density increases. In addition, the economies of density decline with further increases in density. Figure 2–3 summarizes the effects of collection density upon the cost of refuse collection. The top part of the figure shows the relationship between average cost and collection density in the two cities.[m] The bottom part of the figure depicts the estimated relationship between the elasticity of costs with respect to density and collection density itself. Although the cities provide very different types of service, the density elasticities are similar. A 1 percent increase in collection density reduces collection costs by 0.4 to 0.5 percent when density is two tons per linear mile, the average collection density in New Haven; however, a 1 percent increase in collection density decreases collection costs by only about 0.3 percent when density is six tons per linear mile, the average collection density in Hartford. In short, the intracity analysis provides evidence of significant economies of density in refuse collection.

INTERCITY COST FUNCTIONS

A major advantage of the intracity analysis is that it permits a controlled test of the effect of density and hauling distance on

with more than five dwelling units. Collection densities are, on average, three times as large in Hartford as in New Haven. Hartford provides roll-out service for residential collection, while in New Haven curbside service is performed. These differences are reflected in the total costs of collection operations in the two cities and in the costs incurred per hour spent collecting refuse.

[m]Average collection cost is considerably higher in Hartford than in New Haven at comparable densities, in part because of the additional services provided in Hartford and in part because of the fact that the hauling distances are almost twice as long in Hartford as in New Haven. In addition, the treatment of missing data and the use of unweighted regressions may have introduced distortions which make comparison of the level of cost in the two cities inappropriate.

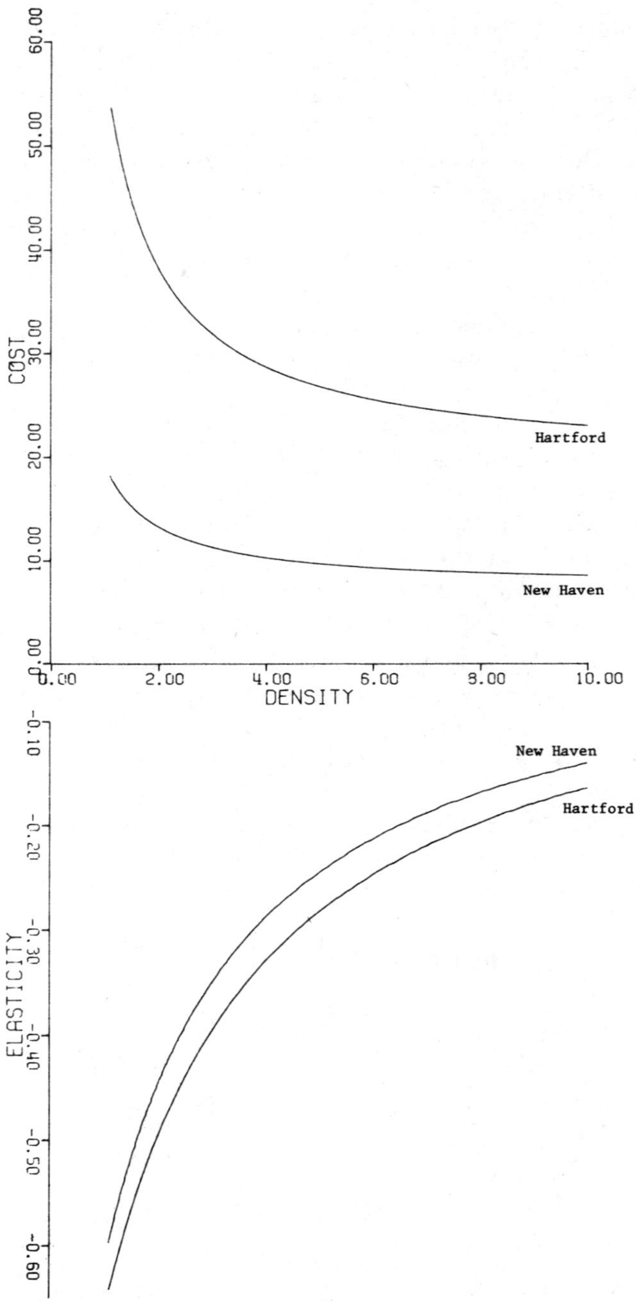

Figure 2-3. Average Costs and Elasticity of Costs, Versus Density, for Hartford and New Haven

collection costs because variations in service, frequency of collection, factor prices, x-efficiency, and organization are held constant. On the other hand, that lack of variation precludes the estimation of the relation between cost and the type of service, frequency, and organization of collection. In this section we extend the analysis to include these factors; in particular, we present estimates of cost functions, which are based on a sample of Connecticut municipalities that contains variation in service, frequency, and organization.

The analysis is based on an extension of the simple model applied previously. The major components of average collection cost are (1) labor and capital costs per unit time, which depend upon factor prices and the amounts of labor and capital used as inputs; (2) pick-up time per ton, which depends upon tonnage density, frequency of collection, type of service, and type of containers used; and (3) hauling time per ton, which depends on haul mileage and the number of tons per load. In addition, variation in the degree of x-efficiency and in the scale of the operation may cause variation in cost. Generalizing the simple model presented in equation (2–3) to account for these factors gives:[n]

$$AC = [c_L\ (w, L) + c_K\ (r, K)]\ [\frac{T_p}{Q}\ (SV, CN, D\ (F), F) \qquad (2\text{-}10)$$

$$+ \frac{T_h}{N}\ (M_h) \bullet \frac{N}{Q}]$$

where:

AC = average cost
w, r = the wage and rental rates
L, K = amount of labor and capital used per unit time
SV = type of service
F = frequency of collection
D = density
CN = type of container used

In its full generality, estimation of equation (2–10) would require

[n]Because frequency of collection affects pickup time in two ways, by reducing effective density and by increasing the amount of stopping and starting, half-empty containers, etc., per ton, it enters the model twice, through density and as a separate explanatory variable.

very detailed information for a large sample of jurisdictions providing refuse collection services. No central organization or agency collects the detailed data needed. Nor is such information even maintained internally by most cities and towns providing or monitoring collection services. Consequently some simplification of equation (2–10) is required before the cost function can be estimated empirically.

First, because separate measures of cost per unit time and collection time per ton are not available, a single equation is estimated with average cost as the dependent variable. Thus, the equation estimated is not directly comparable to the intracity cost functions.[o]

Second, a number of variables are not included simply because they are not available. The type of container used, the average size of load, and haul mileage do not appear for this reason.[p] The first two can be assumed roughly constant across towns; haul mileage does vary, however, creating a potential bias.

Third, because pickup mileage is not available across towns, we were forced to use tons per *square* mile as the measure of density instead of the more appropriate tons per *pickup* mile, used in the intracity analyses.

Given all these simplifications, the model to be estimated is given by:

$$AC = \beta_0 + \beta_1 SV + \beta_2 \left(\frac{1}{D}\right) + \beta_3 F \tag{2–11}$$

where:

AC = average cost
SV = type of service (back-of-the-house or curbside)[q]
D = effective density[r]
F = frequency of pickups

[o]If inputs were used in fixed proportions (i.e., if all collection used three-man crews), and if input prices were constant across towns, then the two models would be directly comparable. However, crew size probably does vary across cities.

[p]Haul mileage was available for a subsample of municipalities. In regressions not reported here, it was consistently insignificant and often had the wrong sign, so it has been excluded from the results reported here.

[q]For those few towns that provide an intermediate type of service this variable takes on a value of 0.5.

[r]Because doubling the frequency of collection is equivalent to halving the density, both measures were adjusted by dividing by the annual frequency of

One additional variable, scale, is included in some of the regressions. While we see no *a priori* reason to believe that there are economies of scale except at very low levels of output, previous research has been interested in this question.[5] To test empirically for the existence of economies of scale, a measure of the number of units served and the total tonnage collected have been included arbitrarily in some of the regressions as measures of scale.[s]

Even with these rather strong simplifying assumptions, assembling data proved to be a difficult task. Information had to be collected by surveying each town, and collection practices and administrative procedures vary from town to town. Consequently, a great deal of effort was expended in obtaining comparable information across towns. Appendixes B and C describe these data collection efforts in some detail.

Three separate samples were collected, consisting of towns with municipal, contract, and private organization of refuse collection. For each of these samples of Connecticut jurisdictions, we estimate two variants of the cost function presented in equation (2–11). In one equation, the dependent variable is annual cost per dwelling unit served; in the other, annual cost per ton collected. The few previous analyses have used cost per unit as the dependent variable, largely because such information is easier to collect. The cost per unit regressions are estimated for comparison with these earlier studies and also because estimates of the tonnage of refuse collected in the sample municipalities are not always reliable. Table 2–7 presents a comparison of the cost per unit regressions for the three samples of Connecticut towns.[t]

For towns providing municipal collection, the regressions explain some 46 percent of the variation in cost per dwelling unit, and the coefficients have the expected signs and magnitudes. Compared to curbside service, back-of-the-house service is significantly more expensive, costing roughly $11 a year more. Twice-a-week collection

---

collection, to obtain effective density. For example, in a town with twice-a-week collection, this measure of effective density would be half as great as an identical town with once-a-week collection.

[s]In separate regressions, scale was measured in three ways: (1) by the number of units (tons) served; (2) the number of units (tons) and its square; (3) the square root of the number of units (tons). In all cases the alternative specifications provided results similar to those reported in Tables 2-10 and 2-11 where alternative (3) is reported.

[t]Means and standard deviations of all variables are reported in Appendix D.

Table 2-7. Intercity Cost Function Estimates: Cost per Dwelling Unit (dependent variable: average cost per dwelling unit)[a]

	Municipal		*Contract*		*Private*
Service (back-of-house = 1)	11.91 (3.4)	11.40 (3.1)	2.82 (0.7)	2.92 (0.7)	14.19 (7.8)
Reciprocal of effective unit density[b]	2.19 (0.2)	5.90 (0.5)	6.41 (0.9)	6.07 (0.8)	−3.44 (1.8)
Annual collection frequency (collection per year)	0.127 (2.2)	0.127 (2.2)	−0.028 (0.3)	−0.032 (0.4)	0.297 (7.5)
Scale (Number of dwelling units served)	—	0.023 (0.7)	—	−0.013 (0.3)	—
Constant	11.52	8.59	19.74	20.93	16.69
R^2	0.46	0.47	0.08	0.08	0.66
Number of observations	23	23	21	21	84

[a] *t*-ratios in parentheses.

[b] Effective unit density is defined as number of dwelling units per square mile divided by annual collection frequency.

is also significantly more expensive than once-a-week service, adding approximately $6 per unit to average cost. As expected, there is no evidence of economies of scale. Contrary to expectation, however, the estimated effect of density upon collection cost is quite small. Although the coefficient has the expected sign in both equations, it is statistically insignificant.

For those cities providing contract service, the estimated regressions of cost per dwelling unit are less satisfactory. The regressions explain very little of the variation in cost per unit and the level of significance is quite low. While the coefficients of service and density consistently have the proper sign, frequency of collection actually has the wrong sign. In short, the cost-per-unit regressions for the contract sample provide little support for the theoretical model.

For the sample of towns providing private service, the overall explanatory power of the regression equations is quite good. Roughly two-thirds of the variation in cost per unit is explained by the regression equation. The level of service is highly significant in the equation; the additional cost of back-of-the-house service is estimated to be approximately $14 per unit per year. Frequency of collection also strongly affects costs; shifting from once-a-week to twice-a-week collection increases costs by $15 per unit per year. The density coefficient, on the other hand, has the wrong sign and is statistically insignificant. This is hardly surprising since, for the sample of towns with private collection, the density measure applies to the town rather than to the firm.

The failure to detect the economies of density in the results presented in Table 2–7 could be due to measurement problems. Two factors are not accounted for in the regressions using cost per dwelling unit as the dependent variable. First, variations in wages may be positively correlated with the level of urbanization, and hence density. If this were the case, higher density would be associated with higher cost. Second, as noted in Chapter 1, the tonnage generated per dwelling unit may be positively correlated with the level of urbanization, and hence density. In an attempt to deal with these factors, Table 2–8 presents regressions, with cost per ton adjusted for wage differences as the dependent variable. Because the tonnage figures are often estimates and the wage adjustment is rather crude,[u] there is undoubtedly considerable measurement error in this dependent variable.

For the sample of towns with municipally provided service, the overall explanatory power of the equation is seen to be lower after these adjustments, probably because tonnage collected was estimated in many towns (rather than weighed, as in New Haven and Hartford). Density is still statistically insignificant and it has the wrong sign when scale is included in the equation.

For those towns providing collection by contract, the regression results improve slightly but remain statistically insignificant. Back-of-the-house service increases collection costs by $3 per ton. Twice-a-week collection increases average cost approximately $5 per ton. The equation also suggests that there are economies of density, although the coefficients are not statistically significant. In short, all variables have the anticipated effect on average costs. The low level of statistical significance may be due both to the small sample size and to the poor quality of the tonnage estimates.

[u]The methods used to standardize cost for differences in wage rates is described in Appendix C.

Table 2-8. Intercity Cost Function Estimates: Cost per Ton (dependent variable: cost per ton).[a,b]

	Municipal		*Contract*	
Service (back-of-house = 1)	6.63 (2.0)	8.17 (2.4)	3.74 (0.7)	3.41 (0.7)
Reciprocal of effective tonnage density[c]	8.43 (0.6)	- 3.28 (0.2)	21.43 (1.3)	26.34 (1.5)
Annual collection frequency (collections per year)	0.093 (1.7)	0.103 (1.9)	0.097 (0.7)	0.107 (0.8)
Scale (total tonnage collected)	—	0.033 (1.4)	—	0.560 (1.1)
Constant	8.52	13.27	3.97	-2.02
R^2	0.25	0.32	0.26	0.34
Number of observations	23	23	15	15

[a]*t*-ratios in parentheses.

[b]For a discussion of the wage adjustment procedure, see Appendix C.

[c]Effective tonnage density is defined as total tonnage collected per square mile divided by annual collection frequency.

Summary of Intercity Results

The generally low statistical power of the intercity regression equations is frankly disappointing, especially since a considerable effort was made to obtain appropriate measures of the cost variables. The estimated coefficients themselves, however, are generally consistent across the three samples and with theoretical expectations, and, except for the coefficient of density, they are of reasonable magnitude. The cost of back-of-the-house collection is roughly 40 percent higher than that of curbside service. Twice-a-week collection is roughly 50 percent more expensive than once-a-week collection. Density is always insignificant in these regressions, however. The point estimates suggest that while there are economies of density in refuse collection, the magnitude of these economies is quite small. At the mean values of the variables in the two samples, the elasticity of average cost with respect to collection density is estimated to be -0.019 and -0.120, respectively, for municipal and contract collection.

It is difficult to reconcile these low estimates of economies of density with the large and highly significant estimates obtained in the analyses of collection costs in both New Haven and Hartford. One possible source of this discrepancy is the difference in the measure of density used in the different analyses. Of necessity, the three analyses which use cities as the units of observation measure density as tons per square mile, while the two intracity analyses use a more appropriate measure of density, tons per pickup mile. In fact, in more dense cities and towns, block sizes are generally smaller than in less dense cities and towns. Thus the elasticity of average cost with respect to tons per *square* mile will underestimate the true elasticity with respect to tons per *pickup* mile.[v]

A second possible explanation of the difference between the two sets of results is based on the possibility of offsetting diseconomies of low density by changing the input mix. As density declines, more time is spent driving between stops, and less actually emptying containers. Since driving is idle time for the laborers, at some point costs are reduced by shifting from three- to two-man crews. At low enough densities, lowest cost may be attained with only a driver who empties containers himself. Since the intracity analyses are limited to three-man crews, this substitution of capital for labor at lower densities cannot be observed. This argument implies that the actual economies of density are less than those observed in the intracity analyses.

A third possible explanation of the discrepancy between the intercity and intracity results is related to the nature of the cities and towns which are served by municipal, contract, and private organiza-

[v]To show this, let:

d_a = tons per square mile
d_r = tons per pickup mile
c = cost per ton
N_{xy} = the elasticity of x with respect to y

In the limit,

$$N_{cd_a} = \frac{N_{cd_r}}{N_{d_a d_r}}$$

Since the area increases more than proportionately with the perimeter, $N_{d_a d_r}$ is greater than one. Consequently, the elasticity of cost per ton with respect to tons per square mile will underestimate the elasticity with respect to tons per pickup mile.

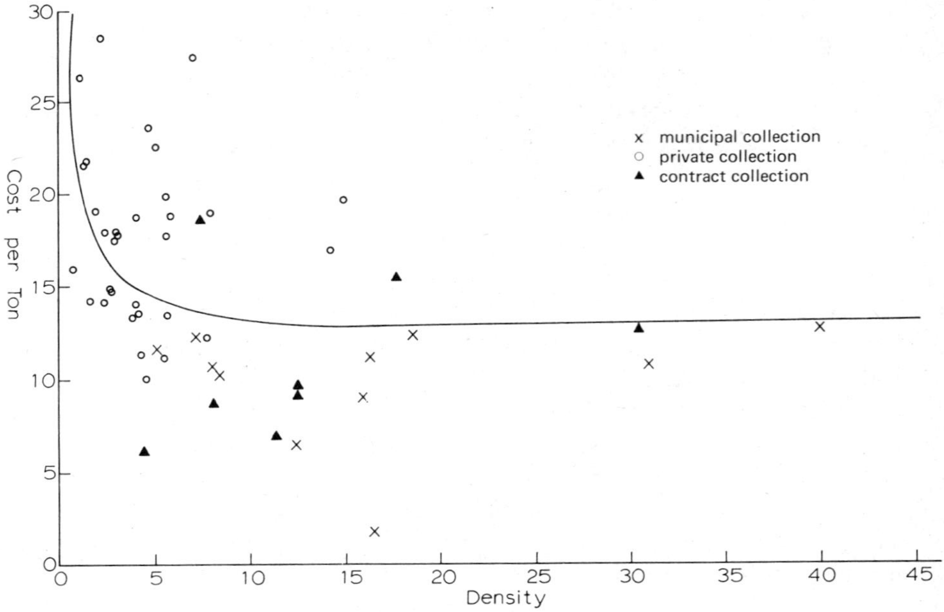

Figure 2–4. Average Cost of Once-a-Week Curbside Service Versus Density

tion. Figure 2–4, which presents a plot of the average cost of once-a-week, curbside service versus density for all three samples combined, illustrates the problem: variation in collection density is not independent of the type of organization. As the figure indicates, an inverse relationship does exist between average cost and density; in fact, the reciprocal specification of density fits the pooled data rather well.[w] But because average cost is also correlated with organization, stratifying the sample by type of organization makes it more difficult to detect economies of density.

Nevertheless, because of the different methods of gathering data for the three samples, because of the different methods of computing the dependent variables from the raw data, and because of the statistical difficulty of separating the effects of organization and density, we have presented the estimates of the model for the three separately without pooling them, because that provides the strongest

[w]The functional relationship depicted in Figure 2–4 is based upon the regression $C/Q = 12.86 + 7.69(1/D)$, where the t-ratio for the estimated coefficient of density is 3.3.

test of the economies-of-density hypothesis. Further discussion of the difficulty of separating the effect of organization and density is deferred to Chapter 3.

NOTES

1. American Public Works Association, *Refuse Collection Practices*, 3rd ed.; (Chicago: Public Administration Services, 1966) p. 79.

2. National Academy of Sciences—National Research Council, *Waste Management and Control: Report to the Federal Council for Science and Technology*, Publication 1400 (Washington, 1966), p. 177. Others have made similar observations. Gueron, in her review of solid waste handling, states that "the paucity of data on all aspects of solid waste problems is evident throughout the literature." See Judith M. Gueron, "The Economics of Solid Waste Handling and Government Intervention," in *Public Prices for Public Products*, ed. Selma J. Mushkin (Washington, D.C.: The Urban Institute, 1972), p. 186. Dennis Young's recent analysis of refuse collection goes further and suggests that "a careful empirical analysis of costs and other variables would be extremely useful," but he also cautions that "such a study will not be easy." See Dennis Young, *How Shall We Collect the Garbage?* (Washington, D.C.: The Urban Institute, 1973), p. 8.

3. Werner Z. Hirsch, "Cost Functions of an Urban Government Service: Refuse Collection," *Review of Economics and Statistics* 47 (February 1965): 87-93. More primitive analyses were also undertaken by H.M. Kitchen and by Robert Clarke, et al. See H.M. Kitchen, *A Statistical Estimation of an Operating Cost Function for Municipal Refuse Collection* (Ottawa: Trent University, 1973); Robert M. Clark, et al., "Cost of Residential Solid Waste Collection," *Journal of the Sanitary Engineering Division*, Proceedings of the American Society of Civil Engineers (October 1972). The few other attempts at collection cost estimates, mainly engineering studies for specific processes, are discussed and criticized in Judith M. Gueron, "The Economics of Solid Waste Handling."

4. Werner Z. Hirsch, "The Supply of Urban Public Services" in *Issues in Urban Economics*, ed. Harvey S. Perloff and Lowden Wingo (Baltimore: Johns Hopkins Press, 1968), p. 508.

5. Werner Z. Hirsch, "Cost Functions of an Urban Government Service: Refuse Collection."

Chapter Three

The Organization of Collection

The single most important policy choice that must be made in any sector of the economy is the mode of organization. Once this choice is made, other issues must be faced, but no other single decision has such a fundamental impact on the performance of a sector. Refuse collection is no exception. Because the output is reasonably well-defined and because several different types of organization are actually encountered, refuse collection provides an especially good opportunity to analyze the effect of the organization of a sector on its economic performance.

There are many different types of organization, ranging along a continuum from private markets to municipal ownership and operation:

Private. The government leaves the provision of refuse collection services to the free market, imposing only minimal health regulations.

Franchise. The government, by franchising private collectors, limits the number of firms competing in the market.

Regulated. The city permits a monopoly but regulates the prices it is allowed to charge and the type of service it provides.

Contract. The government specifies the level of service and contracts with a private firm to provide that service.

Nonprofit. The municipality creates a nonprofit refuse collection agency which is independent of the city government.

Municipal. The municipality provides the service itself through a public agency.

In addition, several combinations are possible such as nonprofit and private.

This chapter analyzes the organization of refuse collection in three sections. The first discusses existing organizational practices using crude data for the nation as a whole and more refined data for the state of Connecticut. The second examines the arguments for government intervention in refuse collection in light of the available empirical evidence. The third, and in some ways the most important section, discusses the effect of organization on the cost of collection using data gathered for our sample of Connecticut towns. The first two sections provide some insight into the positive question of what determines the organization of refuse collection, while the third section provides some evidence on the normative question of what type of collection is best in terms of economic efficiency.

EXISTING ORGANIZATIONAL PRACTICES

As with other aspects of refuse collection, very little systematic information is available about present or past residential refuse collection practices in the United States. The existing survey information, which is biased toward larger cities and includes non-residential collection, is shown in Table 3–1. Approximately two-thirds of the nation's cities have some form of municipal collection service. Since residential refuse is more likely to be collected by the city than industrial or commercial refuse, we infer that in about two-thirds of the cities some municipal agency provides residential collection services. The remaining one-third of the cities are divided between contract and private organization, with a larger proportion providing service through a municipal contract.

Since the samples reported in Table 3–1 do not cover the same cities, it is difficult to infer the trends in collection practices over time. The two surveys conducted by the American Public Works Association (APWA), which are probably the most nearly comparable, suggest that the proportion of cities with contract organization is increasing at the expense of municipal organization. However, comparable survey data over several periods must be collected before meaningful comparisons over time can be made.

The organization of collection varies substantially across regions within the United States. Table 3–2 shows that the percent of cities with municipal organization ranges from a low of 30 percent in the

Table 3-1. The Organization of Refuse Collection in the U.S.

	1955[a]	*1964*[a]	*1968*[b] *Stone*[c]	*1968 National Survey*[d]
I. Municipal				
Municipal only	55%	45%	64%	
Municipal & private	6	15		
Municipal & contract	8	3		
Municipal, contract & private	3	2	8	
Total	72%	65%	72%	65%
II. Contract				
Contract only	15%	18%		
Contract & private	2	5		
Total	17%	23%		
III. Private only	11%	13%		
Total contract and/or private	28%	36%	28%	35%
Number of cities	893	995	234	2,660[e]

[a] American Public Works Association, Committee on Solid Wastes, *Refuse Collection Practices* (3rd ed., Chicago: Public Administration Service, 1966).

[b] Samples are not necessarily comparable.

[c] Ralph Stone and Company, Inc., Engineers, *A Study of Solid Waste Collection Systems Comparing One-Man with Multi-Man Crews* (Public Health Service Publication No. 1892; Washington, D.C.: Bureau of Solid Waste Management, 1969).

[d] Anton J. Muhich, Albert J. Klee, and Paul W. Britton, *Preliminary Data Analysis: 1968 National Survey of Community Solid Waste Practices* (U.S. Department of Health Education & Welfare; Public Health Service, Consumer Protection and Environmental Health Service, Environmental Control Administration; Cincinnati; 1968).

[e] Only towns over 2,500 population.

Northeast and North Central regions to a high of 75 percent in the South Atlantic region. There is also wide variation in the percentage of household refuse collected by city agencies. Under private organization, individuals have the option of taking their refuse to the disposal site themselves. The percentage of household refuse "collected" in this way is quite large and varies across regions.[a]

[a] The sample upon which Table 3-2 is based includes many more small towns than the other national surveys. Moreover, the percentages in Table 3-2 are calculated by weighting by cities rather than population. Together these two factors account for the small proportion of household refuse collected by cities and the large proportion collected by individuals.

Table 3-2. Type of Organization, by Region, for the United States.[a]

Region	Percent of Household Refuse (Volume) Collected[b] by: City	Private Collector	Individuals	Total	Number of Cities	Percent of cities indicating municipal operation of refuse collection.[c]
I Northeast	23.53%	49.34%	26.88%	100.0%	155	30.40%
II North Atlantic	10.70	56.37	32.24	100.0	2,933	35.96
III Mid-Atlantic	52.15	25.52	21.64	100.0	510	59.91
VI South Atlantic	68.97	13.96	16.19	100.0	203	75.00
V Mid-West	35.48	50.71	12.41	100.0	421	45.93
VI North Central	18.78	26.00	54.45	100.0	423	30.67
VII South Central	55.99	13.30	30.05	100.0	727	60.24
VIII Mountain	33.19	31.68	34.63	100.0	185	40.33
IX West	28.19	53.33	17.82	100.0	702	24.60
Entire U.S.	26.29%	43.49%	29.48%	100.0%	6,259	40.90%

[a] Anton J. Muhich, Albert J. Klee, and Paul W. Britton, *Preliminary Data Analysis, 1968 National Survey of Community Solid Waste Practices* (U.S. Department of Health, Education, and Welfare; Public Health Service, Consumer Protection and Environmental Health Service; Environmental Control Administration; Cincinnati; 1968).

[b] Response to Question No. 14, weighted by number of cities.

[c] Response to Question No. 4.

Organizational practices also vary with city size. Three surveys, summarized in Table 3–3, indicate that municipal organization is more common, and private organization less common, in larger cities. In towns with populations of less than 2,500, not only is private organization more likely but households themselves "collect" almost half of the refuse generated. According to the APWA survey, the choice of contract organization is less common is larger cities, although the differences are not as marked. In short, as city size increases, government intervention in the supply of collection services increases significantly, and the type of intervention is more likely to be municipal ownership and operation of refuse collection. These results suggest that intervention in refuse collection is associated with urbanization.

There also appears to be a relation between government intervention and residential density. Table 3–4 shows the median density of towns with various types or organization. Although the differences are small, density increases as organization changes from private to contract to municipal.

Table 3–4 also indicates that there is a rather marked association between median income and type of organization. Low-income cities appear much more likely to have municipal organization, while high-income cities are more likely to have contract or private organization.

While our sample of Connecticut towns is smaller, the better-quality data permits a more careful examination of these relationships than is possible using the larger national surveys. Table 3–5 shows the relation between organization and city size, density, and income. Part A indicates rather clearly that as city size increases, government intervention is more likely. Over 80 percent of Connecticut towns with populations of less than 10,000 choose private organization, but this percentage decreases to 8 percent for cities of over 50,000 population. The fraction of towns contracting with private firms increases up to a population of 50,000, beyond which it declines. For the very largest towns, municipal collection is the dominant form of organization. The percent of towns with municipal organization increases rather sharply from a low of 1.7 percent among the very smallest towns to 83 percent among the largest towns. In short, as city size increases, the dominant form of organization shifts from private to contract and then to municipal, the same pattern observed in the national data.

Table 3-3. Type of Organization, by City Size, for the United States.

A. Survey by American Public Works of 1964[a]

	Population Class (thousands of persons)							
Type of Collection	*5-10*	*10-25*	*25-50*	*50-100*	*100-1000*	*1000+*	*Not Stated*	*Total*
I. Municipal								
Municipal only	47.2%	47.9%	46.0%	35.0%	37.2%	—	27.3%	45.0%
Municipal & private	6.1	12.8	13.6	22.7	44.0	80.0%	27.3	15.0
Municipal & contract	.4	2.8	6.0	8.2	1.3	—	9.0	3.0
Municipal, contract & private	1.2	1.1	3.0	1.0	1.3	20.0	—	2.0
All Municipal	54.9	64.6	68.6	66.9	83.8	100.0	63.6	65.0
II. Contract:								
Contract only	22.0	19.5	15.1	13.4	9.0	—	9.0	18.0
Contract & private	3.3	3.9	6.0	7.2	3.8	—	—	4.0
All contract	25.3	23.4	21.1	20.6	12.8	—	9.0	22.0
III. Private only	20.0	12.0	10.1	12.4	3.8	—	27.3	13.0
Total	100.0	100.0	100.0	100.0	100.0	100.0	100.0	100.0
Number of Cities	246	359	199	97	78	5	11	995

Type of Collection	Population Class: Rural (0-2500 population)	Urban (2500 population)	All Cities
1. Per cent of cities of each size where city operates collection	23.8%	64.5%	40.9%
2. Percent of household refuse collected by:			
Public agency	9.9	48.6	26.3
Private collector	43.7	42.8	43.5
Individuals	45.8	7.8	29.5
3. Number of Cities	3,454	2,663	6,259

C. 1968 Survey by Ralph Stone and Company, Inc.[c]

Type of Collection	Population Class (thousands of persons)			
	10-100	*100-500*	*500 or more*	*Total*
I. Municipal				
Municipal only	60.0%	78.9%	76.0%	64.5%
Municipal & private	6.7	7.9	17.6	7.6
All Municipal	66.7	86.8	93.6	72.1
II. Private only	33.0	13.2	5.9	28.0
III. Number of cities	179	38	17	234

[a]American Public Works Association, Committee on Solid Wastes, *Refuse Collection Practices* (3rd ed.; Chicago: Public Administration Service, 1966).

[b]Anton J. Muhich, Albert J. Klee, and Paul W. Britton, *Preliminary Data Analysis: 1968 National Survey of Community Solid Waste Practices* (U.S. Department of Health, Education, & Welfare, Public Health Service, Consumer Protection and Environmental Health Service; Environmental Control Administration; Cincinnati, 1968). Questions No. 4 and 14.

[c]Ralph Stone & Company, Inc., Engineers, *A Study of Solid Waste Collection Systems Comparing One-Man with Multi-Man Crews* (Public Health Service Publication No. 1892; Washington, D.C.: Bureau of Solid Waste Management, 1969).

Table 3-4. Type of Organization by Income and Density for U.S. Cities of Population over 25,000

	Median Density (population per square mile)		*Median Income (dollars)*		*Number of Cities*	
I. Municipal	4,382		6,023		646	
Municipal only		4100		5970		446
Municipal & private		5024		6091		151
Municipal & contract		4913		6506		33
Municipal, contract & private		5092		5862		16
II. Contract	4,306		6,771		219	
Contract only		4048		6781		175
Contract & private		5334		6735		44
Average						
III. Private only	4,192		7,018		130	

Source: Dennis Young, *How Shall We Collect the Garbage?* (Washington, D.C.: The Urban Institute, 1972, p. 10). Note that averages of the medians are reported because medians for the groups were not reported in the original source.

Part B of Table 3–5 shows the way in which organization varies with residential density. As density increases, the probability of government intervention increases markedly. Among the lowest density towns, only 3 percent have some form of government intervention, whereas all of the highest density towns have government intervention. Among towns with either contract or municipal organization, the proportion with contract collection decreases with increases in density. There is a strong correlation between both the level and form of government intervention and density of the city.

Part C of Table 3–5 shows the same relation between income and organization in Connecticut that was observed for the United States as a whole. Government intervention varies inversely with median income. Moreover, among those towns with some form of government intervention, the proportion contracting increases consistently with income.

In summary, in Connecticut and in the nation as a whole, refuse collection practices vary significantly with city characteristics such as size, density, and income. Government intervention in general, and municipal provision in particular, are much more likely as city size and density increase and as income decreases.

Table 3-5. Percent of Connecticut Towns Choosing Municipal, Contract, or Private Organization, by Town Characteristics

A. Organization by City Size

Population	*Municipal*	*Contract*	*Private*	*Number of Towns*
0 - 10,000	1.7%	16.9%	81.4%	59
10,000 - 25,000	23.1	28.2	48.7	39
25,000 - 50,000	26.7	46.7	26.7	15
50,000 and over	83.3	8.3	8.3	12
Total	19.0%	23.0%	57.9%	125

B. Organization by Density

Housing Units per Square Mile	*Municipal*	*Contract*	*Private*	*Number of Towns*
0 - 100	0.0%	3.4%	96.6%	29
100 - 200	11.5	15.4	73.1	26
200 - 500	25.0	54.2	20.8	24
500 and more	57.7	42.3	0.0	26
Total	22.6%	27.4%	50.0%	105

C. Organization by Income

Median Income	*Municipal*	*Contract*	*Private*	*Number of Towns*
0 - $ 9,000	33.3%	27.8%	38.9%	18
9,000 - 11,000	26.2	28.6	45.9	42
11,000 - 13,000	18.9	29.7	51.4	37
13,000 and over	0.0	7.1	92.9	14
Total	21.6%	26.1%	52.3%	111

REASONS FOR GOVERNMENT INTERVENTION

What explains this pattern of government intervention? The standard economic reasons for government intervention in a sector include externalities, redistribution of income, and economies of scale. For the delivery of local services there are several additional reasons, including economies of density, tax advantages, contracting cost, administrative bias in cost comparisons, and other considerations. In this section we examine these reasons for intervention in light of the empirical evidence. Examination of this evidence should provide some information on the desirability of government intervention, and it may suggest why the choice of organization appears to vary systematically with certain city characteristics.

Externalities

One motive for government intervention in refuse collection is the existence of externalities, such as health hazards and unsightly litter, which would result if refuse were not collected. In an attempt to develop measures of the effectiveness of refuse collection, a recent Urban Institute study catalogs many of the important externalities and measures one, the litter condition, in several neighborhoods in the city of Washington, D.C.[1] The analysis, based on a block-by-block inspection, indicates substantial variations in the amount of litter across neighborhoods.

We have noted a strong positive relation between government intervention in refuse collection and the density of the city. A simple explanation is that, as densities increase, externalities become more bothersome, creating a demand for government intervention in order to minimize their adverse effects. Table 3–6, which matches the Urban Institute's measure of litter condition by block with census data, supports this hypothesis. The table shows rather strikingly that the density in the littered areas is higher than in the clean areas. If this relation holds for other externalities, such as rats, insects, and smell, both within and across cities, then externalities may explain the observed relation between density and the likelihood of government intervention.

It is important to emphasize, however, that even if externalities turn out to be extremely important considerations in the decision to intervene, public refuse collection is not a guaranteed solution, nor is it the only solution to the problem of externalities. Government intervention typically takes the form of providing collection services free of charge and raising revenue through city taxes. But providing service at a zero price does not mean that people will take advantage of it because they must still expend some effort in order to do so. From the individual's point of view, tossing a candy wrapper on the ground is "cheaper" than carrying it to the litter basket. An alternative solution to the problem of externalities is strict enforcement of existing laws which forbid littering and unsightly or hazardous accumulation of refuse. Since such laws are required even when the service is provided at a zero price, providing the service at a zero price may not significantly reduce the negative externalities. In short, while the externalities associated with refuse provide a strong argument for government provision of refuse collection, especially in

Table 3-6. Litter Ratings by Density, for a Portion of Washington, D.C.

Litter Condition	*Total Population*	*Total Year-Round Housing*	*Population per Square Mile*	*Housing Units per Square Mile*
Clean	14,989	5,303	12,714	4,498
Slightly littered	22,296	7,659	32,274	11,087
Littered	3,778	1,038	29,944	8,227

Sources: Litter ratings and area are from Louis H. Blair and Alfred I. Schwartz, *How Clean is Our City?* (The Urban Institute, Washington, D.C.: 1972), pp. 54, 55. Population and number of housing units are from U.S. Department of Commerce, Bureau of the Census, *1970 Census of Population and Housing* (Washington, D.C.: U.S. Government Printing Office, 1972).

dense areas, they do not appear sufficient by themselves to explain the high correlation between government intervention and density.

Redistribution

A second potential justification for government intervention in refuse collection is that it benefits the poor. A crude estimate of the magnitude of the redistribution possible by government intervention is the ratio of refuse cost per dwelling unit to family income. For our sample of Connecticut towns, the average of this ratio is quite small, about 0.3 percent. It is difficult to believe that the redistribution of such a small percentage of income could motivate government intervention.

However, the evidence indicating substantial variation across neighborhoods in the amount of litter present suggests a more subtle distributive argument for intervention. Since litter increases with density, it undoubtedly decreases with income. If, more generally, the incidence of externalities varies substantially with the income, the poor may well bear the brunt of the burden of externalities. It is conceivable that intervention could be justified in order to redistribute the burden of these externalities. Nonetheless, although government intervention undoubtedly helps the poor, a case for intervention based on redistribution alone is not persuasive.

Economies of Scale

When scale economies are large, competition among firms is inefficient because the scale of operation implied by a large number of competing firms is too small. If substantial economies of scale exist, some form of monopoly is required to take advantage of them, necessitating some type of government intervention.

In refuse collection there are two potential sources of economies of scale, arising from indivisibilities and the law of large numbers. First, if certain types of equipment can only be purchased in rather large, discrete "lumps," then—up to some point—applying additional units of other inputs will not require the addition of another lump of the indivisible equipment. As a result, costs will decline with increases in output, as the cost of the indivisible input is spread over more units of output. For example, for a small operation with a single truck, capital costs will fall until that truck is fully utilized. Second, the law of large numbers suggests that if certain types of equipment are used only on an irregular and unpredictable basis, then increased size may increase the utilization of that equipment. For example, if backup trucks are required in the event of breakdown, then a large firm will require fewer backup trucks per ton collected than a smaller firm. This would also cause average costs to decline with increases in scale.

A priori, neither type of economy appears to be important for residential refuse collection as currently practiced. As described in more detail in Chapter 2, the production process is extremely labor-intensive, the only specialized equipment necessary is a compactor truck, and the administration is not complex. Consequently, once a city reaches a population of 2,000, so that it is large enough to utilize one truck fully,[b] economies due to indivisibilities are unlikely to be important for residential refuse collection. Neither is the law of large numbers likely to account for substantial economies of scale. While trucks do break down, any economies from having backup trucks on hand are likely to be quite small. In short, except for very small operations, there is little reason to expect to find economies of scale in refuse collection.

The empirical evidence supports this *a priori* judgment. Table 3–7 shows the size distribution of private residential collection firms in the United States. Over half the firms have three or fewer trucks, and 80 percent have fewer than ten trucks. If substantial economies of scale existed, it is difficult to see how such a large number of small

[b]A fully utilized truck collects roughly 15 tons per day or 780 tons per year. At 0.4 tons per person per year, a population of 2,000 will fully utilize one truck.

Table 3-7. Size Distribution of Private Residential Collection Firms in the United States.

Size of Contractor (Number of Trucks)	*1*	*2-3*	*4-5*	*6-9*	*10-19*	*20-49*	*50+*
All firms	22.9%	31.7%	15.2%	11.4%	11.6%	4.9%	2.3%
Private	24.8	33.2	16.2	11.3	10.1	2.5	1.9
Subcontract	10.9	23.0	10.9	14.9	21.3	13.2	5.7

Source: Applied Management Sciences, Inc., *The Private Sector in Solid Waste Management* (U.S. Environmental Protection Agency, 1973), p. 65.

firms could continue to exist. Moreover, regression results reported in Chapter 2 provide no evidence of economies of scale. The coefficient of the scale variable was never statistically significant, and more than half the time the sign of the coefficient indicated that there are diseconomies rather than economies of scale. In summary, there is little *a priori* reason to expect economies of scale, and there is no empirical evidence that they exist.

Economies of Density

In contrast to economies of scale, there are strong *a priori* reasons to expect economies of density. As density increases, the distance between pickup units decreases and the collection cost per ton correspondingly declines. Thus, if there are several collectors competing along the same route, the effective density will be lower for each, and the cost of collection will be higher.

The empirical evidence on economies of density is consistent with this expectation. In the intercity analyses, reported in Chapter 2, where the samples are small and the density measurements are far from perfect, the coefficient of density is usually statistically insignificant, but the signs of the estimated coefficients suggest that economies of density exist. In the two intracity analyses, where the density measures are better, the estimated cost functions show significant economies of density. Specifically, they indicate that a 10 percent increase in tonnage density reduces average cost by up to 5 percent.

In short, both the *a priori* arguments and the empirical evidence indicate that residential refuse collection is subject to economies of density. These economies support some form of public intervention in refuse collection. The reasons are analogous to those involving economies of scale; a monopoly over a well-defined spatial area is required to minimize average cost, but under such a monopoly private firms will charge a price in excess of marginal cost. Government intervention of some sort is required to ensure marginal cost pricing. Of course, the same problems which apply to government regulation or provision in sectors where there are economies of scale also apply to those sectors where there are economies of density.

Contracting Cost

Each type of organization has different contracting costs associated with it. Under private organization each individual incurs the cost of searching for a private collector, making an agreement with him, and enforcing it. These costs can be high relative to the cost of service, especially if collectors are unreliable. Under contract organization, city officials are responsible for receiving bids, negotiating contracts, and enforcing compliance. Finally, under municipal organization, the contracting functions are replaced with administrative functions including negotiating with unions, purchasing and arranging for the maintenance of trucks, dealing with citizen complaints, and so on.

There may be economies of scale in contracting which would explain the difference in organization by city size.[c] Thus, in a small town the total contracting cost incurred by individuals under private collection may be less than the contracting cost the city would incur under contract organization. Similarly, a public works official with other responsibilites may be able to handle contracting and enforcement in an intermediate size town, but a full-time person may be required to administer a municipal agency. If such administrators are not divisible, then only larger cities would be expected to choose municipal organization. Unfortunately, we can only speculate about the magnitude of contracting costs.

Tax Advantages

The tax laws distort the choice of organization by providing tax advantages to governmental units which are not available to private firms. There are two different types of tax distortions.

First, because cities are exempt from paying federal and state taxes, municipal provision is "cheaper" from the point of view of

[c]We are grateful to Bernard Saffran for suggesting this.

Table 3-8. Estimated Tax Savings of Municipal Provision for Hartford, Connecticut[a]

Tax	*Annual Savings[b]*	*Percent of Annual Cost*
Sales tax	$ 8,400	0.5%
Federal & state income tax	64,300	3.8
Federal & state gas tax	15,100	1.0
Vehicle registration	6,700	0.4
Total	$94,500	5.7%

[a]For details of these calculations see Appendix B.

[b]These figures do not include local property tax payments, which are noted in Table 3-9, part C.

city residents than either contract or private provision. For example, since city governments are not required to pay federal and state gasoline taxes, their operating costs will be lower than those for otherwise identical private firms. Of course, this "saving" is not a saving of real resources, but is simply an intergovernmental transfer. Nonetheless, from the point of view of city residents, it is a "real" saving in dollars which they do not have to pay. Similar tax savings exist because cities do not have to pay state sales taxes, federal and state corporate income taxes, or state vehicle registration fees. Table 3-8 estimates these savings for the city of Hartford. The total tax saving amounts to $94,500 per year or 5.7 percent of the annual cost of refuse collection. So this tax incentive to choose municipal over contract or private organization is a rather substantial one.

Second, taxes distort the choice of organization because of the federal income tax deductibility of state and local taxes. Because a city resident can deduct property taxes from his income for federal income tax purposes, he does not have to pay the full resource cost of municipally financed refuse collection. For example, if he is in a 33 percent marginal tax bracket, then for every dollar he contributes to city revenues, he receives, in effect, a rebate from the federal government of 33¢. This provides a rather potent incentive to finance refuse collection through local taxes.[d] In principle this incentive encourages tax financing, but not the choice of organization. In practice, however, this incentive affects the organization choice

[d]Estimates of the magnitude of this incentive are presented in Chapter 5 on the financing of refuse collection.

because only contract and municipal organization can be financed through property taxes. Thus, while federal deductibility provides an incentive for choosing contract or municipal over private or franchised provision, it does not distort the choice between contract and municipal provision. The incentive to finance refuse collection through taxes can be quite substantial.[e] Whether government intervention to take advantage of the tax incentive is desirable depends upon one's point of view. From the point of view of the nation as a whole, local governments should consider only the real resource cost of collection and not the artificial tax incentives, but from the point of view of city residents, any saving is desirable, whether it is a real resource saving or simply a tax saving.

Finally, we should note that although the magnitude of the first tax distortion might be quite different for other municipal services, both distortions apply in principle to virtually all types of municipal services. The importance of these tax distortions appears to have been overlooked both by public officials and by economists. Further analysis of these tax incentives and their impact is clearly warranted.

Biased Cost Comparisons

As is well known to students of public finance, municipal budget and expenditure data do not represent the full economic cost of resources devoted to any particular function. Although municipalities typically maintain a separate budget and expenditure account for refuse collection, these figures seldom include all of the costs that are properly attributed to that service.

In the first place, there is a limit to the extent of "program budgeting" undertaken at the local level. For example, in many local accounts, all fringe benefits—a sizeable figure in a labor-intensive service such as refuse collection—are reported as a single separate item, rather than being allocated to operating agencies. Similarly, expenditures for motor vehicle maintenance may be reported in the accounts for the public works department or elsewhere, depending on local practice.

Second, few local governments engage in systematic capital budgeting. Thus the cost of capital equipment, such as refuse collec-

[e]Indeed we have heard a second-hand account of a local Connecticut politician who argued for municipal collection on grounds that even though it is slightly more expensive than private collection, the federal income tax saving more than outweighs this additional expense. Whether this account is in fact correct or not, it is clear that city politicians might well make such an argument.

tion vehicles, garage and office space, is rarely amortized over its useful life, and is often reported in a separate "capital expenditure" account.

Finally, municipalities virtually never attempt to estimate, much less to attribute to operating functions, the opportunity costs of their activities. These opportunity costs include the alternative yield of municipal investment in capital equipment.

Although local officials presumably realize that a simple comparison of municipal accounts for refuse collection with contract bids is biased, it is not obvious that they recognize the magnitude of the bias in such comparisons.

For the cost analyses reported in Chapter 2 and discussed further in this chapter, it was necessary to obtain reasonably accurate and consistent estimates of the true resource costs of refuse collection for a sample of Connecticut cities. A detailed analysis of budgets and expenditures for Hartford, discussed in detail in Appendix B, is summarized in Table 3–9. The table indicates that the municipal accounts underestimated the resources devoted to refuse collection

Table 3-9. Possible Bias in Cost Comparisons.[a]

A. Budget and Estimated Cost of Refuse Collection for Hartford, Connecticut, Fiscal year 1972–1973.

Category	*Budget Figure*	*Estimated Cost*
Direct labor	$1,111,855	$1,259,785
Administration	12,130	28,869
Vehicle maintenance & operation	0	254,217
Capital costs	61,870	
Depreciation	0	68,914
Interest cost	0	59,743
Other	0	526
	$1,185,855	$1,672,054

B. Budget and Estimated Cost for 18 Connecticut Cities

Average budget figure:	$348,101
Average cost estimate	$451,264
Percent underestimate	
Average	22.9%
Average percent	20.6%

C. Estimated Understatement of Municipal Cost Due to Foregone Property Tax for Hartford, Connecticut, Fiscal year 1972–1973

Property Tax	$ 39,100	2.3% of budget

[a]See Appendix B for methodology and assumptions.

in Hartford by $486,000, or by almost 41 percent. Of course, a substantial reason for this discrepancy is the fact that fringe benefits and vehicle maintenance are reported in separate accounts. It is unlikely that public officials would not take these costs into account when making cost comparisons. Yet if decision-makers took all budget items properly into account, and only neglected interest cost—which appears nowhere in the local accounts—they would still underestimate the true cost of collection operations by some 3.6 percent. Thus it is likely that decisionmakers underestimate the true cost of municipally provided collection services when comparing the costs of alternative forms of organization. Such biases are understandable, given local accounting practices.

An additional source of bias in cost comparisons of different organizational structures arises from local tax laws. For example, when a comparison is made between the municipal cost of collection and the contract cost, the municipality is not likely to include the foregone property taxes in its estimate of cost. Although these foregone taxes are not true savings (since the municipality would pay the tax to itself), it is clear that the contracting firm must include property taxes in its bid. Part C of Table 3–9 shows that this underestimate for the city of Hartford amounts to about 2.3 percent of annual collection cost, again a substantial distortion.

So there are two possible sources of bias in comparing the costs of alternative organizational choices: (1) arising from accounting practices; and (2) local tax considerations. Both biases lead to underestimates of municipal costs by an amount that could be sufficient to alter the choice of organization. Whether policymakers are in fact misled by these distortions in their evaluations of alternatives is an unresolved behavioral question.

Other Considerations

While the economic arguments presented above may be important, the organizational decision is made through the political process. Several additional explanations of the observed pattern of organization choice exist. One is simply that ideology varies with income. Rich citizens are more likely to be opposed to government intervention of any sort, regardless of tax savings and economies of density.[f] A second explanation is that because the political process in

[f]We are grateful to Charles Gilbert for suggesting this explanation.

large cities is more complex, politicians need more resources with which to bargain. A large refuse collection agency can provide needed patronage.

Summary

The economic arguments for government intervention in residential refuse collection based on income redistribution and economies of scale are not persuasive. However, three other arguments make a persuasive case for such intervention: the existence of substantial externalities, economies of density, and artificial tax incentives. In addition, there are other possible explanations of the observed pattern of organization choice. Government accounting could easily mislead policymakers in evaluating the cost of alternative types of organization, biasing the choice in favor of municipal collection. Contracting cost under each type of organization may vary with city size, explaining the variation in organization with city size. Finally, political considerations provide an alternative set of explanations of the observed pattern of choice of organization.

While the economic case for government intervention appears to be quite strong, the analysis so far has not considered differences in the cost of collection, which may be systematically related to the type of organization. If private collection is significantly more efficient than either municipal or contract collection, then the gains from efficiency might more than outweigh any losses due to externalities, foregone economies of density, or foregone tax savings. The following section analyzes the costs of collection by type of organization and provides comparisons of the costs of municipal, contract, and privately provided service.

ORGANIZATION AND COST

There are four sources of difference in average cost which one might expect to be associated with organization. First, the existence of economies of density increases the cost of private provision compared to contract or municipal provision. If several competing trucks cover the same route, cost per ton is higher than if a single truck covered that route.

Second, there may be differences in factor prices inherent in different forms of organization. Since refuse collection is labor-intensive, labor costs are the important consideration. Since private firms are typically small, and a rise in wages may force the private firm out of business, there is some reason to expect that private firms might make greater efforts to economize on wage bills than city governments, whose guaranteed existence and political visibility may make them less able to resist wage demands. Provided bidding is truly competitive, contracting firms might also be expected to pay lower wages than city governments.

Third, the efficiency[g] of operation may vary by organization. How efficiency varies with respect to organization depends upon differences in organizational behavior. In the absence of well-developed models of organizational behavior, economists are usually quick to emphasize the virtues of profit in motivating, and competition in enforcing, efficiency in production and distribution. Since municipalities are neither motivated by profits nor policed by direct competition,[h] municipal provision is seen as less likely to be efficient than private or contract provision, so long as these are truly competitive.

Finally, however, private and contract organization need not be truly competitive. Competitive private markets can evolve into cartels or monopolies, and under contract organization, the choice of contractor is subject to political pressure or even graft. Not only does the existence of monopoly lead to pricing in excess of marginal cost and to monopoly profits, but it may also weaken the profit incentive for efficiency. Thus, monopoly can be expected to lead to higher cost. To the extent that monopoly exists in either private or contract provision, either one may be more expensive than municipal provision.

While theoretical analysis is both necessary and instructive, the organization choice cannot be based solely on *a priori* reasoning. The way in which costs vary with type of organization can only be determined by comparing actual behavior. *A priori* reasoning may be prone to comparing the virtues of ideal competition with the vices of government bureaucracy and insulation. The appropriate comparison is between municipal, contract, and private organization as they actually function. Our sample of Connecticut towns allows such an empirical comparison.

[g]In this context, efficiency is used rather broadly to refer to the choice of inputs used, to getting the greatest output from the inputs used (x-efficiency), and to the search for new ways of reducing costs.

[h]In the longer run, residential mobility between municipalities may enforce competitive behavior.

Cost Comparisons

The simplest evidence that can be examined is the collection cost per dwelling unit for each type of service, and frequency of collection under alternative forms of organization. Part A of Table 3–10 shows that the average cost per dwelling unit for private collection is considerably higher than that for municipal collection, which in turn is somewhat more expensive than contract collection.

These differences in cost per dwelling unit could arise from regional variations in wages or from variations in the tonnage generated by each household. Part B of Table 3–10 presents a comparison of costs—deflated for regional wage differences—per ton of refuse collected.[i] Even after these adjustments are made, however, substantial differences in cost exist among the various types of organization.

This comparison omits one very important consideration: the effect of density upon average cost. As noted earlier, the type of organization varies systematically with density. Since increases in density lead to decreases in cost, the high cost of private collection may be simply the result of the lower residential density in areas where private organization prevails. Figure 3–1, repeated from Chapter 2, illustrates the difficulty of separating the effect of organization from the effect of density on average cost by presenting the cost of once-a-week curbside service plotted against density. At first glance, it appears that average cost is inversely related to density and that much of the variation in average cost could be explained by density. However, closer examination suggests an alternative explanation of the differences in average cost: average cost varies with the type of organization, but the type of organization chosen happens to be correlated with density.

Regression analysis has been used to test this hypothesis statistically by pooling the municipal, contract, and private samples. In order to control for the type of service and frequency of collection, the sample has been stratified into four subsamples, one for each combination of service and frequency. Part A of Table 3–11 shows that before controlling for organization, there is a significant inverse relation between average cost and density in each of the four subsamples. After controlling for the effect of organization by including dummy variables for private and contract organization, however, the

[i]Since tonnage data are not directly available for the private sample, tonnage figures were estimated using the refuse generation equations developed in Chapter 4. The details of the tonnage imputation and the procedure for adjusting for wage differences are noted in Appendix C.

Table 3-10. Average Annual Cost, by Service, Frequency, and Organization.

A. Cost per Dwelling Unit

Type of Organization	All Types of Service and Frequency	Once-a-Week[a]		Twice-a-Week[a]	
		Back-of-House[b]	Curbside	Back-of-House[b]	Curbside
Private[c]	$42.49 (100)[d]	$45.26 (35)	$30.34 (40)	63.43 (19)	$41.10 (6)
Municipal[e]	23.28 (23)	27.26 (5)	19.08 (12)	38.31 (2)	23.40 (4)
Contract	19.24 (22)	22.71 (3)	17.34 (13)	22.19 (1)	21.52 (5)

B. Cost per Ton[f]

Type of Organization	All Types of Service and Frequency	Once-a-Week		Twice-a-Week	
		Back-of-House	Curbside	Back-of-House	Curbside
Private	$23.50 (86)	$24.92 (30)	$17.69 (33)	32.25 (17)	$23.52 (6)
Municipal	15.36 (23)	18.58 (5)	12.39 (12)	28.77 (2)	16.55 (4)
Contract	12.09 (22)	12.00 (3)	10.55 (13)	17.06 (1)	14.54 (5)

[a]Once-a-week includes annual frequencies of less than 78; twice-a-week, 78 or more.

[b]Back-of-house includes roll-out service.

[c]Note that municipalities appear more than once in the private category because residents have choice of service and frequency.

[d]Number of municipalities in category is in parentheses.

[e]Municipal cost estimates have not been adjusted for differences in tax treatment of municipal governments.

[f]Cost has been deflated for regional wage differences using data from Connecticut Labor Department. Tonnage data were imputed using a refuse generation equation from Chapter 4. See Appendix C for details.

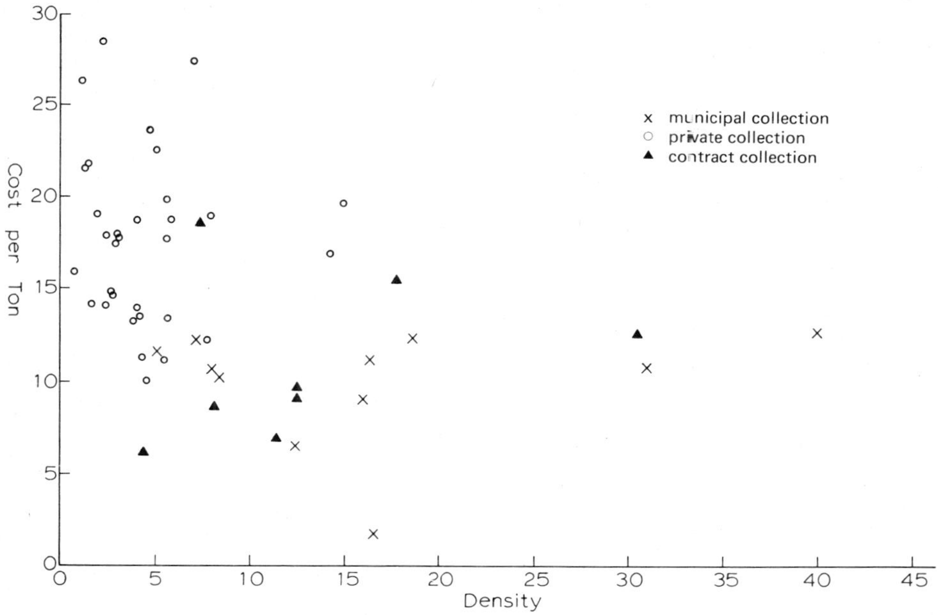

Figure 3-1. Average Cost of Once a Week Curbside Service Versus Density

inverse relation between average cost and density is both less pronounced and less significant, as is shown in Part B of Table 3-11. Organization has a large effect on cost even after controlling for density. The variable representing private organization has a large positive coefficient, which is significant in three of the four subsamples, and the variable representing contract service has a negative coefficient, which is significiant in two of the four subsamples. In short, even after controlling for density, the same story emerges: private collection is most expensive, contract collection is least expensive, and municipal collection is in between.

It is important to examine the magnitude of these differences in cost to see whether they are worth worrying about. Part C of Table 3-11 shows the difference in cost due to organization as a percent of the cost for a typical town under municipal organization.[j] Depending on the service and frequency, private collection is from 25 to 36

[j]Technically, the predicted cost has been calculated using the regression equation, assuming average density and municipal organization. The ratio of the coefficient of the organization dummy variable to the predicted cost of municipal collection provides the estimates of the cost increase due to organization in Part C of Table 3-11.

Table 3-11. Pooled, Intercity Cost Regressions Stratified by Service and Frequency (dependent variable: cost per ton)[a]

A. Cost as a Function of Density

Frequency of collection:	*Once-a-Week*[b]		*Twice-a-Week*[b]	
Service:	*Back-of-House*[c]	*Curbside*	*Back-of-House*[c]	*Curbside*
Reciprocal of tonnage density	10.89	7.69	5.64	12.71
	(2.2)	(3.3)	(1.9)	(3.5)
Constant	19.91	12.86	28.19	15.13
R^2	0.12	0.17	0.16	0.50
Number of observations	37	57	20	15

B. Cost as a Function of Density and Organization

	Once-a-Week[b]		*Twice-a-Week*[b]	
	Back-of-House[c]	*Curbside*	*Back-of-House*[c]	*Curbside*
Reciprocal of tonnage density	2.94	4.14	2.42	5.16
	(0.6)	(1.9)	(1.1)	(0.9)
Private organization	5.22	4.39	8.50	4.38
	(1.8)	(2.8)	(3.2)	(1.2)
Contract organization	– 5.77	– 1.85	– 16.04	– 2.22
	(1.5)	(1.0)	(1.5)	(0.9)
Constant	18.45	11.79	22.61	16.22
R^2	0.35	0.38	0.67	0.63
Number of observations	37	57	20	15

C. Effect of Organization on Cost as a Percent of Predicted Cost under Municipal Provision[d]

	Once-a-Week[b]		*Twice-a-Week*[b]	
Type of Organization	*Back-of-House*[c]	*Curbside*	*Back-of-House*[c]	*Curbside*
Private	+27.1%	+34.0	+36.0	+24.8
Contract	−30.0	−14.3	−25.6	−12.6

[a]Cost per ton is deflated for wage differences and tonnage is imputed using procedures described in Appendix C. *t*-ratios are in parentheses.

[b]Once-a-week service is defined as 78 collections per year or less; twice-a-week as more than 78.

[c]Back-of-house service includes roll-out service.

[d]Numbers shown are the coefficient of the dummy variable for the indicated type of service divided by the predicted cost assuming municipal organization and average density, all times 100.

percent more expensive, and contract collection is from 13 to 30 percent cheaper than municipal collection. These are substantial differences in cost by any standard.

Taken at face value these results suggest that towns should switch from private and municipal to contract provision. This would mean a major change in the organization of refuse collection in the United States since nearly 80 percent of the cities of over 5,000 people have either municipal or private collection.

A Critical Look at the Cost Comparisons

But before recommending so profound a change in collection practices, we should examine our conclusion carefully.[k] In this section we reexamine some of the reasons for differences in cost with organization. Specifically we investigate differences in collection densities and wage rates, evidence of monopoly, differences in the quality of service, in the choices available to consumers, and in administrative cost.

Reduced Density of Private Collection. The reduction in effective density that results from competition along a route is one possible

[k] We have taken pains to eliminate any obvious biases in the underlying data which would explain differences in cost by organization. There were three potential biases in the data used to make the comparisons of cost across types of organization.

The first is failure of municipal costs to include an estimate of foregone property taxes and state and federal taxes which municipalities do not pay. As mentioned earlier in this chapter, that understatement amounts to approximately 5 percent of total costs for the city of Hartford. While municipal costs probably should be adjusted to include these taxes, the magnitude of the adjustment is too small to change the ranking of cost among the three types of organization; indeed, adjusting for property taxes would increase the differential between municipal and contract collection.

A second potential bias exists because the private cost data refer to cost conditions which occur later than the municipal and contract costs. To take account of inflation, firms were asked for their current charges and their charge four years ago. Using their response, the average inflation rate was calculated to be 7.5 percent per year. The private data were collected approximately a year and a half later than the municipal and contract data. They were deflated by 15 percent, a figure that is greater than the average inflation rate, in order to be absolutely certain that there is no inflationary bias in the private cost estimates.

A third potential bias results from the dump fees private firms must sometimes pay, but contracting firms and municipalities do not have to pay. To get some idea of the magnitude of this potential bias, we asked city officials what the dumping charge for private collectors was. Typically there was no charge, or only a minimal fee, for a permit to dump. Three towns charge as much as $600 per truck per year for a permit to dump, but even this fee is small in relation to total cost. For the one town charging $5 per ton, cost was reduced by this amount to insure comparability.

explanation for the high cost of private collection. There is little direct evidence on the number of collectors serving each route. Conversations with town officials revealed that in some towns there is a great deal of overlap among routes, but in others there is no overlap at all. In some cases, these officials provided estimates of the number of collectors serving a town which ranged from two to sixty-three, averaging twelve collectors per town. Unfortunately, the number of collectors serving a town is not a good measure of the number of firms competing on each route. Nonetheless, thirteen of the fifty-two towns where information was available reported five or fewer collectors serving the entire town. Based on this evidence and our casual observation, we guess that there are rarely more than a few collectors serving each route.[1]

Although direct evidence is unavailable, we can make some indirect inferences concerning the importance of reduced effective densities resulting from competition. The regression estimates of the effect of density on cost per ton indicate that there would have to be from three to eight competing collectors on each route to explain the difference between the cost of private and municipal collection. Thus it seems unlikely that the higher cost of private collection results entirely from the reduction in effective density due to aggressive competition. Moreover, even if this were the explanation of the higher cost of private collection, the policy implication is still that private collection should be avoided.

Differences in Wage Rates. A second explanation of the differences in cost is that wage rates vary by type of organization. One hypotheses is that municipalities, most of which have organized labor forces, are required to pay higher wages than firms in competitive industries. While this would not explain why the cost of private provision exceeds the cost of municipal provision, it would explain why the cost of municipal provision exceeds the cost of contract provision.

Since private firms are reluctant to make information on costs

[1]Because these estimates include the number of commercial collectors, they probably overestimate the number of residential collectors in each town. Moreover, the number of collectors is highly dependent upon the size of the town; large towns obviously have more collectors, even when the number of competing collectors on each route is the same. Finally, even if there are many collectors, they need not compete along each route.

public, no information on their actual wage scales is available. However, we can investigate whether municipalities pay wages which are in excess of the competitive wage by comparing the wages of refuse collectors with those in the private sector. Table 3–12 compares the starting wages of municipal refuse collectors with the average earnings of laborers and material handlers in the few Connecticut cities where data are available. Because the starting municipal wage is undoubtedly below the average municipal wage, municipal wages are biased downwards, and the comparison thus provides only a weak test.

Table 3–12 shows that in one area, average starting municipal wages exceed the average earnings in the private sector, but that in the other, the average starting municipal wage is substantially lower. In seven of the ten towns for which these data are available, the starting municipal wage is lower than average earnings in the private sector. Thus the table provides little or no evidence that municipal wages exceed the going market wage rate. It therefore appears that wage differences, if they exist, are not sufficient to explain the observed cost differences among different types of organization, at least in the relatively small towns in Connecticut.

Differences in efficiency represent a third source of cost variation by type of organization. The motivation of profits and the discipline of competition are expected to lower the cost of private and contract provision. While this is certainly a possible explanation of the lower cost of contract collection, it appears contrary to the higher cost observed for private collection. For obvious reasons, independent evidence on the efficiency of operation is not easily obtained.

Monopoly. The existence of monopoly is another potential source of differences in cost with type of organization. It may explain the apparent failure of competition to lower the cost of private collection. Although, as Table 3–7 indicates, entry barriers are quite low, there is considerable evidence that monopoly in refuse collection may be a serious problem. Refuse collectors in several areas of the United States are prosecuted for price-fixing under federal antitrust laws. Since the prosecuted cases probably only represent the tip of the iceberg, and since collusion can also take the form of dividing territories rather than fixing prices, these cases suggest that monopoly is a real problem in refuse collection.

Table 3-12. Comparison of Starting Wages for Refuse Collectors and Regional Average Earnings of Laborers, 1973.

Area	*Average Hourly Earnings of Laborers & Material Handlers in All Industries, 1973*[a]	*Starting Wage of Refuse Collectors in Municipalities*[b]			*No. of Municipalities with Wages Greater than the Average Market Wage*	*Number of Towns*
		Average	*Minimum*	*Maximum*		
Norwich-New London-Groton	\$3.22	\$2.89	\$2.50	\$3.42	1	5
Stamford-Norwalk-Stratford	3.35	3.43	2.84	4.05	2	5

[a]U.S. Department of Labor, Bureau of Labor Statistics, *Area Wage Survey* for New Haven, Bridgeport, Norwalk, Stamford, Norwich, Groton, New London, Waterbury, Springfield, Chicopee, Holyoke.

[b]From telephone conversations with officials in respective town halls.

Town ordinances provide some hints that monopoly may be a problem. For example, one ordinance states that:

> Any action that is found. . .to be an attempt by the licensee. . .to obtain an exclusive right to a specific route or to specific customers may subject such licensee to loss or suspension of license.[2]

In another town, collectors are required to furnish route maps and to report rates to the cities, although no regulation is imposed. Finally, in several towns the director of public works is specifically empowered to order a private firm to collect from a resident who is dissatisfied with the service of another collector. While these regulations are not positive evidence of monopoly, they do suggest that cities are concerned at least about the potential of monopoly power.

Conversations with city officials responsible for local collection also revealed concern about monopoly. Several officials reported that in their towns the routes of private collectors do not overlap. Others volunteered that they must exert informal pressure on private collectors to insure that each area is served by at least two collectors. One official reported his private belief that collectors actually fixed prices. Another reported that when the town attempted to switch from private to contract collection, no company would bid for the contract. While this evidence is all very casual and does not apply to all towns, we were left with the distinct impression that collusion is a problem of some consequence.[3]

One would like to obtain more objective evidence on the existence of monopolies, such as profit rates or concentration ratios. Needless to say, small family-held companies do not make public reports of their profits, nor are they willing to release them informally. Since the standard objective measures of monopoly power are not available, we have been forced to rely on cruder measures. From conversations with town officials we learned whether routes overlapped in a city, whether there was a collectors' association, or whether there was some other indication of monopoly. The towns providing private service have been divided into two groups, those with any of these three indications of monopoly and all others. A comparison of the average cost of collection for these two groups, shown in Table 3–13, indicates that costs are higher when there is some reported evidence of monopoly. Regressions which control for the effect of other variables are reported in Table 3–14. These regressions show that the cost of private collection is higher in towns where there is some evidence

Table 3-13. Average Cost per Unit, by Frequency of Collection and by Evidence of Monopoly[a]

	Frequency of Collection	
	Once-a-Week	*Twice-a-Week*
Evidence of monopoly	$56.04 (2)	$67.80 (5)
Others	44.61 (33)	61.87 (14)
All firms	45.26 (35)	63.43 (19)

[a]Only back-of-the-house service is included in this table because in the private sample all towns where curbside service is offered show no evidence of monopoly. Evidence of monopoly is defined as absence of route overlap, existence of a collectors association, or comments of town officials suggesting a lack of competition.

For many towns, no information of the existence of monopoly was available; they are treated as having no evidence of monopoly.

Number of towns shown in parentheses.

Table 3-14. Cost Regressions Including Evidence of Monopoly as an Explanatory Variable for the Private Sample.[a] (dependent variable: cost per ton).[b]

	Excluding Monopoly Measure[c]	*Including Monopoly Measure[c]*
Reciprocal of tonnage density	2.56	4.10
	(1.0)	(1.4)
Annual collection frequency	0.139	0.124
	(4.8)	(3.9)
Evidence of monopoly (Yes = 1)		2.68
		(1.2)
Constant	16.54	16.67
R^2	0.39	0.41
Number of observations	46	46

[a]Only municipalities with back-of-the-house service are included in this sample because no evidence of monopoly was found among towns where curbside service was offered.

[b]Cost per ton is adjusted for wage variation, and tonnage is imputed as explained in Appendix C.

[c]t-ratios in parentheses.

of monopoly, but the higher cost is not statistically significant. The low significance is not surprising, given the crude measurement of "monopoly."

In summary, the evidence lends qualitative support to the hypothesis that the high cost of private collection is partly due to noncompetitive practices. There are numerous policy alternatives which might serve to eliminate this problem, the most obvious being to switch to collection by competitive contract. But alternatives also exist. The strict enforcement of antimonopoly laws might have a beneficial effect. Alternatively, as Dennis Young suggests, the creation of a nonprofit, independent organization to compete with the private collectors might force them to lower prices. Of course, these alternatives are only designed to deal with the anticompetitive practices but will do nothing to deal with the economies of density which are foregone under competition.

Quality of Service. We have not accounted for a number of important dimensions of quality which could not be measured. It is possible that differences in the amount of remaining litter, noise, or in other services are related to the type of organization. Private collection costs may be higher simply because private firms provide higher quality service than municipalities or contracting firms. While this is possible, it does not seem sufficient as an explanation since the cost differential is quite substantial.[m]

Consumer Sovereignty. Another important consideration that is ignored in the cost comparison is the lack of choice offered under contract and municipal organization. Where provision is private, an individual can choose the type of service and frequency of collection he wishes to buy, but this consumer sovereignty is lost when the government intervenes and selects a single type of service for all residents. This is a significant cost of existing modes of government intervention. It should be emphasized, however, that the trade-off between this freedom of choice and the cost of collection is a rather steep one.[n] Moreover, as discussed in Chapter 5, consumer sovereignty is possible under contract and municipal organization.

[m]Moreover, there is no casual evidence that private quality is better; indeed, town regulations give the impression that the quality of the service may be a problem of private service. For example, one town regulation states: "Refuse collectors shall leave their customers' containers in a neat, upright position with lids replaced and off the road . . . [and] shall clean refuse that may have spilled . . ." Our conversations with town officials revealed some dissatisfaction with the dependability and littering of private collectors. However, a careful study would be needed to verify this extremely casual evidence.

[n]In addition, it is not altogether obvious that government intervention by contract provision or by provision through a municipal agency must necessarily restrict all consumer choice, although it may generally restrict consumer choice

Contracting Cost. Also ignored in the cost comparison is the cost of contracting itself. Under private collection, each individual must make his own arrangements with a private collector, pay his bills monthly, and arrange for a new collector if his present one should go out of business.[o] While these costs are probably not terribly great, they are certainly not zero. If anything, the contracting costs of private collection are an additional argument for municipal organization at least under current pricing policies.

There are also contracting and enforcement costs incurred by municipalities which contract with the private firm. These costs include those of preparing specifications and negotiation, and—perhaps most important—accepting complaints and enforcing the contract. While the magnitude of this additional cost is not trivial, it is difficult to believe that these costs are large enough to outweigh the estimated advantage of contract collection.

While there are other noncost considerations that are important and may be overriding in some cases, none appears to be sufficiently persuasive to outweigh the advantages of cost.

CONCLUSION

A major finding of this comparative cost analysis is that the type of organization of refuse collection really is significant. The cost of refuse collection differs substantially among towns with private, municipal, and contract forms of organization. Private collection seems to be more expensive than municipal collection, which is in turn more expensive than contract collection. Although they are subject to some qualifications, our examination of these results leads us to conclude that local governments should intervene in the collection of refuse to take advantage of the economies of density and to eliminate any adverse effect of monopoly pricing, but they should

---

about service frequency. For example, in at least one town in our sample, curbside collection service is provided by a town contractor to all households whose head is younger than age 65, while back-of-the-house service is provided to "senior citizens". In cities in other parts of the country, different levels of service are provided at different charges at the resident's choice. These possibilities for allowing citizens to register demands for alternative service levels are considered in more detail in Chapter 5.

[o]The private firms incur some costs of billing made necessary by the market form of organization. These costs are included in the data used to compare the different types of organization, and, while they are probably not substantial, they do serve to raise the cost of private collection.

not provide the service themselves. Rather, they should take advantage of the profit motive and discipline afforded by competitive bidding.[4]

NOTES

1. L.H. Blair and A.I. Schwartz, *How Clean is our City?* (Washington, D.C.: The Urban Institute, 1972).

2. "Notice to Refuse Collectors," Darien, Connecticut, November 1973.

3. In several parts of the United States, judicial concern with monopoly in refuse collection has existed for some time. For example, in the New York metropolitan area court actions were instituted as early as 1957 to drive racketeers from the garbage collection industry. In November 1974, court actions were instigated by the New York State Attorney General's Office charging restraint of trade under state law. See Ralph Blumenthal, "Garbage Has Been a Dirty Racket for a Long Time," *New York Times*, January 5, 1975, Section 4, Page 6, Column 4.

4. For a suggestion of the method in which such contracts should be put up for bid, see Dennis Young, *How Shall We Collect the Garbage?* (Washington D.C.: The Urban Institute, 1973), pp. 31-35, 73-76. He suggests a number of specific policies designed to take maximum advantage of competitive bidding.

Chapter Four

The Generation of Refuse and the Demand for Collection

Many of the decisions made by local governments and regional authorities require estimates of the amount of refuse generated in particular jurisdictions. Decisions about the appropriate size and location of the disposal facility or the feasibility of recycling operations, for example, are sensitive to projections of future refuse generation. Operational decisions about the efficient design of collection routes and about the timing of investment in collection vehicles depend upon estimates of the amount of refuse generated, its variance, and its spatial distribution within a collection area. In short, better understanding of the determinants of refuse generation could be used to improve refuse collection and disposal planning and to increase the efficiency of service provision.

In addition to better understanding refuse generation, estimates of the demand for different types of service and frequencies of refuse collection are needed for other decisions. For example, estimates of the price elasticity of demand for different types of service are needed to choose the type of service for an entire city or to estimate the benefits of financing schemes which permit residents a choice of service. Useful as such estimates would be, we are limited by data availability to consideration of the generation of refuse.

CONCEPTUAL FRAMEWORK

The amount of refuse generated by an individual or household can be expected to depend on each of the following factors:

Income. Because consumption of goods and services increases with income, we expect the amount of refuse generated, and the resulting demand for collection services, to increase with income. In addition, there is reason to expect that at higher incomes, households may substitute disposal for labor, generating more refuse. For example, because the time of higher income families is more valuable in the marketplace, they may economize on the time spent in meal preparation by consuming more convenience foods which involve more packaging.

Household Size and Demographic Composition. Household refuse generation can be expected to be responsive to variations in household size, with larger households generating more refuse than smaller households. Since consumption patterns also vary with age and with other demographic characteristics, differences in household refuse generation are to be expected. For example, casual empiricism suggests that children and teenagers generate more refuse than other segments of the population, perhaps because they consume more packaged goods and fewer services than adults.

Price. If the price per ton of refuse collected varied, we might expect this variation to affect the amount of refuse generated. At higher prices consumers could rearrange their configurations of commodities chosen, (even at the same incomes and commodity prices) to reduce the level of refuse generated. Moreover, at higher prices, consumers would have an incentive to dispose of refuse in other ways, including disposers, recycling, burning, and so on. While this is not a change in the amount of refuse generated, it will be perceived as such by city collection agencies. It should be emphasized that, since household demand for collection services is derived from demand for goods and services, and since the cost of virtually all goods is large relative to the cost of the disposal of its associated refuse, the price elasticity of demand for collection upon the tonnage generated is likely to be quite small.

Service and Frequency. In many communities residents do not have a choice of the type of service provided since that decision is made by the municipal government. In these cases, variation in service acts like a price by making it more or less difficult to have refuse collected. For example, if there were neither private nor public refuse collectors, the "price" of refuse collection faced by a household would be the necessity of making regular trips to the dump. In this case, we might expect households to generate less solid waste than if collection services were provided at the rear of the house, daily, at zero marginal cost. Again, the magnitude of this effect is likely to be quite small.

Other Factors. In addition to these factors affecting the variation in the amount of refuse generated at the household level, we may expect geographic, climatic, and other regional differences to affect refuse generation. For example, lawn clippings and brush will be more prevalent in medium density areas than in urban centers or in rural areas.

EXISTING EVIDENCE

Practically no information currently exists on the determinants of the demand for collection services. Judith Gueron's recent survey of solid waste handling states: "There is little information beyond an indication that pricing can affect the amount collected. . . . There is *no* quantitative evidence on income elasticity. . . . The empirical evidence on the importance of the effect of any variables is minimal."[1]

One systematic attempt to analyze factors affecting the generation of residential refuse and the demand for service has been prepared for the Environmental Protection Agency (EPA).[2] During a period of three weeks the refuse collected from each structure along three collection routes in a city was weighed. The published analysis includes tabulations of the total and per capita refuse generated on each route for single detached houses, small multiple dwelling units, and apartments. Also presented are tabulations indicating the relationship between the per capita refuse collected in a structure and the number of occupants. Based upon these tabulations, the study comes to several contradictory conclusions and cautions against making estimates of refuse generation on an unadjusted per capita basis.

By reanalyzing some of the data gathered for the EPA study, it is possible to get some general insight into the relationship between economic variables and the demand for collection services by households. Table 4–1 presents averages of certain characteristics of the three collection routes (A, B, and C) and the average refuse generated

Table 4–1. Selected Characteristics of Housing and Waste Generation along Routes Servicing Low-Income Population, Cincinnati, 1972

Route	*Median Assessed Value of Single Detached Units*	*Percent of Families with Income Greater than $3,600*	*Percent of Single Detached Units with Greater than 50 feet frontage*	*Average Household Size in Single Detached Units*	*Refuse per Capita per Week (mean pounds)*	*Refuse per Unit per Week (mean pounds)*
A	$11,440	76.2%	50%	4.0	16.07	64.28
B	10,280	59.2	41	4.4	11.69	51.44
C	6,660	60.4	15	4.5	11.51	51.80

Source: George R. Davidson, Jr., *A Study of Residential Solid Waste Generated in Low-Income Areas* (Cincinnati: U.S. Environmental Protection Agency, 1972), from Tables 1, 2, 4, 6, 7 (sample sizes not given).

along each route. Comparison of the routes suggests that higher income households generate more solid waste per capita than lower income households.

DEMAND FUNCTIONS

This suggestive but hardly definitive evidence appears to be the only empirical analysis of the determinants of the demand for refuse collection. It should be possible, however, to estimate a demand model empirically using either routes in one city or a cross-section of cities.

For jurisdictions which provide municipal or contract service, typically only one level of service is provided, and it is financed from general revenues. Thus, within a jurisdiction there is little or no variation in service levels, and the price for collection faced by households is essentially zero. If information could be obtained on the refuse generated by households within a jurisdiction, a model of the following form could be estimated:

$$Q_i = f_i\,(Y, A) \qquad (4\text{–}1)$$

where:

Q_i = tonnage generated if provided service type i
Y = income
A = vector of socioeconomic variables
i = type of service and frequency of collection

Because the frequency and service level are constant within one jurisdiction, a relatively pure test can be performed of the effect of income, family size, and other demographic factors, on the amount of refuse generated.

If similar information could be obtained for a sample of jurisdictions which provide different levels of service, it would be possible to investigate whether refuse generation varies when different types of service are provided, even if unit prices are not charged.

$$Q = g(Y, A, SV, F) \qquad (4\text{–}2)$$

where:

SV = type of service
F = frequency

The remainder of this chapter is devoted to developing empirical estimates of the effect of the economic factors upon refuse generation in urban areas. The two analyses presented below are based on data collected in Connecticut. The first analysis relies upon information gathered about the characteristics of collection routes and the populations served within a single city, New Haven. In this *intracity* analysis, demand functions similar to Equation (4–1) are estimated. From this analysis we hope to learn something about the income elasticity of demand for solid waste collection at a particular level of service.

The second analysis is based upon a slightly larger sample, which also includes observations on several cities, towns, and special districts in Connecticut which provide collection services either through a city agency or by government contract. In this *intercity* analysis, demand functions similar to equation (4–2) are estimated. The second analysis increases the range of variation in the variables important to refuse generation and permits an investigation of the service elasticity of demand for collection.

The Intracity Analysis

As explained in Appendix A, each truckload of refuse collected by the New Haven Public Works Department is weighed at the municipal incinerator, providing accurate estimates of the tonnage collected along each route for an entire year. By matching the collection routes to census blocks, the socioeconomic information available through the U.S. Census at the tract and block level could also be compiled for each route. The resulting sample includes the annual residential tonnage generated on thirty-three routes in 1972 and socioeconomic characteristics of the residents served.[a]

Table 4–2 reports the means and standard deviations of selected characteristics of New Haven's collection routes based on this sample. The estimated average refuse generation per household in the city is 1.11 tons per year, or about 2.1 pounds of refuse per capita per day, well below the national average of 3.0. It should be noted that in terms of total tonnage collected throughout the year, there is little variation across the collection routes. On average, each route

[a]The population served by municipal collection includes the residents of single detached housing units and the residents of multifamily structures containing fewer than five dwelling units; in the aggregate, residential refuse collection is provided to about 68 percent of the dwelling units in New Haven.

Table 4-2. Means and Standard Deviations of Selected Characteristics of New Haven Collection Routes (33 observations)

Variable	*Mean*	*Standard Deviation*
Tons collected per year	1,420.50	118.31
Tons per dwelling unit	1.11	0.32
Tons per person	0.38	0.09
Median household income	9,312.80	2,164.90
Per capita income	4,210.60	2,006.30
Persons per household	2.84	0.34
Children per household	0.83	0.28
Aged per household	0.36	0.10
Proportion children	0.28	0.08
Proportion aged	0.13	0.04
Units served	1,014.00	143.44
Population served	3,950.40	1,332.50
Households served	1,498.90	722.60

generates some 1,420 tons of refuse per year, the standard deviation is only 118 tons, and the coefficient of variation is, therefore, only 0.08.

Since we do not have strong *a priori* reasons for choosing among alternative specifications of the model, we report the results of several alternatives. In separate regressions the dependent variables are tons per household and tons per capita. In addition, we report the results using different combinations of the explanatory variables for both the linear and the logarithmic form. Examination of the consistency of the results across the different specifications thus supplements the normal statistical tests in assessing the importance of specific variables.

Table 4–3 presents the regression results using the linear specifications. With one exception, the regressions are significant by conventional statistical criteria, and the coefficient estimates are stable as variables are added or dropped. The signs of the coefficients are consistent with expectations and their magnitudes are reasonable.

Income has a strong positive effect on the tonnage generated in the demand functions estimated on a per capita or per household basis. Household income by itself explains 22 percent of the variation in households' generation of refuse; income by itself explains 56 percent of the variation in tonnage generated in the per capita regressions. Income remains statistically significant and the coefficient varies little as other variables are introduced into the regression. The

Table 4-3. Regression Estimates of Refuse Generation Based on New Haven Sample (linear form) (number of observations: 33 routes)[a]

Dependent Variable	*Tons per Household*				*Tons per Person*			
Regression number	*1*	2	3	4	5	6	7	8
Income (thousands of dollars)[b]	0.069		0.064	0.054	0.034		0.032	0.028
	(2.89)		(3.82)	(2.71)	(6.20)		(5.54)	(4.96)
Persons per household		0.631	0.619	0.616		0.093	0.030	0.039
		(4.87)	(5.71)	(5.73)		(2.03)	(0.80)	(1.23)
Density				−0.597				−0.388
				(0.91)				(2.43)
Constant	−0.469	−0.681	−1.226	−1.064	0.241	0.119	0.161	0.206
	(2.06)	(1.84)	(3.64)	(2.79)	(9.48)	(0.90)	(1.73)	(2.33)
R^2	0.218	0.442	0.632	0.643	0.561	0.120	0.573	0.647
Income elasticity (estimated at means)	0.452		0.419	0.353	0.375		0.353	0.309

[a] *t*-ratios in parentheses.

[b] Median household income is used in the tons per household regressions; per capita income is used in the tons per person regressions.

magnitude of the income effect is substantial. The coefficients in regression number 5 imply that, on average, refuse generation would increase by a little more than 136 pounds (18 percent) per person per year along a route where incomes are larger by one standard deviation ($2,006). Thus we observe that normal variations in income lead to substantial differences in the refuse generated across collection routes.

The income elasticities, computed at the means, suggest that a 1 percent change in household income is associated with a 0.31 to 0.45 percent change in the amount of refuse generated.[b] These income elasticity estimates are similar to those obtained by other researchers for the demand for such items as foodstuffs, beverages, newsprint, and transportation services.[3]

Family size has a highly significant positive effect on the demand for collection services. Adding one more person to the household increases household refuse generation by 1,200 pounds. This is a surprisingly large number, because it is some 60 percent above the average per capita generation. In the per capita regressions, when the number of persons per household is included with income, the coefficient is still positive but statistically insignificant. These results suggest that larger families generate more refuse per capita than smaller families. It is possible that larger families with the same income increase the proportion of goods to services in their market basket, leading to a greater per capita generation of refuse.

When gross residential density is also included as an explanatory variable, there is only the weakest evidence of an inverse relationship between refuse generation and density in the household regressions. In contrast, regression number 8 suggests that there is a small but nevertheless significant variation in per capita refuse generated at different density levels. The coefficient implies that per capita demand for collection services increases by about 50 pounds (2.3 percent) per year along a route where gross residential density is lower by one standard deviation.

In addition to the results reported in Table 4–3, we estimated regressions using the additional information on the age distribution of the population instead of persons per household. Neither per

[b]The income elasticities estimated on a per capita basis are slightly lower than those estimated on a per household basis. The larger income elasticity may be due to the difference in the income measures used. Because the income distribution is positively skewed, a 1 percent change in median income is associated with more than a 1 percent change in average income.

capita nor per household refuse generation was found to be significantly related to the age distribution of the populations served.

To test the sensitivity of the results to the linear functional form, the same refuse generation equations were estimated using a multiplicative specification where all variables are expressed as logarithms. The results, shown in Table 4–4, are quite consistent with those for the linear form. In general, the R^2's are larger for the logarithmic specification, and the statistical significance of the individual regression coefficients is at least as great. The income coefficient is highly significant in all specifications of both models. For per capita refuse generation, the constant income elasticities estimated from the logarithmic models, 0.32-0.39, are virtually identical to those estimated from the linear additive models. For the equations estimating refuse generation per household, the income elasticities computed from the logarithmic specification, 0.67-0.82, are substantially higher than those estimated from the linear models. The family size measure is positive and highly significant in the generation per household regressions, though less significant in the per capita generation equations. Again, in the equation predicting per capita refuse generation, there appears to be a significant inverse relationship between residential density and the demand for collection services, but its magnitude is very small. In short, with the exception of the higher income elasticities in the per capita regressions, the conclusions do not change with the logarithmic specification.

The Intercity Analysis

The analysis based on the sample for a single city provides estimates of the income elasticity of demand for refuse collection at a given level of service. However, it provides no evidence about possible effects of service variations on the amount of refuse generated. To explore this hypothesis, we have expanded the sample to include information on an additional 35 Connecticut municipalities which provide residential refuse collection. Tonnage data for these towns were collected through a special survey described in Appendix C. Socioeconomic data were obtained from census publications. Since many of the towns in Connecticut are approximately the size of a collection route in New Haven, pooling the samples is not inappropriate; it should be made clear, however, that roughly half of the sample consists of New Haven routes. Moreover, since some

Table 4-4. Regression Estimates of Refuse Generation Based on New Haven Sample (logarithmic form) (number of observations: 33 routes)[a]

	Tons per Household				Tons per Person			
	1	*2*	*3*	*4*	*5*	*6*	*7*	*8*
Income (dollars)[b]	0.817		0.760	0.671	0.393		0.373	0.323
	(3.12)		(4.64)	(3.09)	(6.90)		(5.93)	(5.14)
Persons per household		1.969	1.904	1.915		0.883	0.204	0.307
		(5.52)	(6.92)	(6.27)		(2.41)	(1.73)	(1.17)
Density				−0.064				−0.161
				(0.63)				(2.30)
Constant	−7.398	−1.982	−8.844	−7.481	−4.220	−1.922	−14.277	−2.450
	(3.10)	(5.34)	(5.88)	(2.77)	(9.04)	(5.03)	(8.89)	(2.71)
R^2	0.245	0.564	0.715	0.719	0.606	0.158	0.613	0.673
Income Elasticity	0.817		0.760	0.671	0.393		0.373	0.323

[a]t-ratios in parentheses.

[b]Median household income is used in the tons per household regression; per capita income is used in the tons per person regressions.

towns do not weigh each load, the tonnage estimates are subject to a great deal of measurement error.

Table 4–5 presents regressions relating refuse generation per household to incomes, family sizes, and gross residential densities for both the linear and the logarithmic forms. Also presented for both the linear and the logarithmic form is a regression that includes service (that is, the location of pickup) as an independent variable and an additional regression which includes collection frequency.[c] The results reported in Table 4–5 must be interpreted with caution; as indicated by the R^2 for each equation, the regressions explain a much smaller proportion of the variance in tonnage per household than those based on the smaller sample of New Haven routes.

Nevertheless, expanding the sample generally does not change the results based on the New Haven sample by itself. The estimates of the income elasticity of demand range between 0.44 and 0.67; household size and density have the same relative effects upon refuse generation.

When the service parameters are included among the explanatory variables, there is no evidence that households generate more refuse if it is collected more frequently. There is some weak indication that households generate more refuse if they are served more conveniently by back-of-the-house collection. The results from the linear regression suggest that households generate an additional 140 pounds of refuse per year if back-of-the-house service is provided instead of curbside service, an increase of 6 percent.[d]

In other regressions, not reported, the per capita generation functions were shown to be quite similar to the household regressions; in still other regressions, neither per capita nor per household refuse generation was found to be significantly related to the age distribution of the population served.

SUMMARY

The empirical analyses presented in this chapter suggest that there is a substantial amount of variation in the amount of refuse generated

[c] Since tonnage and frequency may be jointly determined, these regressions were estimated by two-stage least squares; the results are virtually identical with the ordinary least squares estimates.

[d] Of course, it is not possible to know whether this tiny difference in refuse actually collected results from the nuisance to households of carting 30-gallon garbage cans to the curb, or from the increased possibilities of spillage, littering, and the like, associated with curbside collection.

Table 4-5. Refuse Generation Equations Estimated from Expanded Sample: Tons Per Household (number of observations: 68)[a]

	Linear Form						*Logarithmic Form*					
	1	*2*	*3*	*4*	*5*	*6*	*7*	*8*	*9*	*10*	*11*	*12*
Income (thousands of dollars)	0.101		0.080	0.067	0.079	0.080	0.671		0.462	0.486	0.457	0.448
	(2.36)		(1.20)	(1.44)	(1.81)	(1.80)	(3.25)		(2.36)	(2.19)	(2.33)	(2.27)
Persons per household		0.603	0.439	0.370	0.440	0.456		1.820	1.550	1.431	1.551	1.406
		(2.10)	(1.46)	(1.22)	(1.52)	(1.33)		(4.50)	(3.21)	(3.28)	(3.21)	(2.95)
Residential density $\times 10^4$				−0.118						−0.031		
				(1.09)						(1.78)		
Service (back-of-house = 1)					0.071	0.071					0.181	0.169
					(2.08)	(2.06)					(1.06)	(1.00)
Frequency (collections per year) $\times 10^2$						0.494						0.117
						(0.09)						(0.59)
Constant	0.384	−0.418	−0.705	−0.279	−0.762	−0.856	−5.929	−1.755	−5.703	−5.068	−5.662	−4.911
	(0.89)	(0.49)	(0.82)	(0.30)	(0.91)	(0.64)	(3.13)	(3.97)	(3.31)	(2.65)	(3.29)	(2.29)
R^2	0.079	0.064	0.109	0.125	0.166	0.170[b]	0.140	0.237	0.299	0.315	0.311	0.316[b]
Derived income elasticity	0.648		.513	.430	.507	.513	0.671		0.462	0.436	0.457	0.448

[a] *t*-ratios in parentheses.

[b] Estimated by two-stage least squares: R^2 is from the ordinary least squares regression.

by urban households and thus in their demand for refuse collection services. The average amount of refuse generated per household is about 1.11 tons, but the coefficient of variation is 0.29. Thus, simple extrapolations based upon the number of individuals or households served are unlikely to provide adequate predictions of the demand for refuse collection which are needed for refuse collection and disposal policy planning.

In contrast to simple extrapolation, several alternative specifications of simple demand models indicate that as much as two-thirds of the variation in refuse generation across collection routes can be explained by differences in household incomes and family size. The following example may suggest the magnitude of these systematic variations in the demand for refuse collection services.

A two-person household earning $6,000 per year in New Haven would generate about 800 pounds of solid waste per year; a four-person household earning $12,000 per year would generate about 4,000 pounds of refuse a year. Thus, doubling both income and family size would increase the demand for refuse collection by a factor of five.

The estimates of the income elasticity of demand for refuse collection range from 0.3 to 0.8. Variables other than income and family size do not seem to affect the amount of refuse generated in a significant way. No evidence that the frequency of collection affects household refuse generation was found. There is only weak evidence that more refuse is generated if households are more conveniently served by back-of-house collection. In addition, the results provide only weak support for the hypothesis that residential density affects household generation of refuse. If anything, the results suggest that per capita and per household demand for collection varies inversely with residential density, perhaps reflecting the effects of lawn clippings and the like in less dense residential areas.

The analysis presented in this chapter indicates that projections of the demand for collection and disposal services based upon extrapolation of historical trends or upon estimates of the number of households to be served are likely to be misleading. Changes in the rate of population growth are likely to make the extrapolation of historical trends unreliable; changes in income and family size are likely to make per household or per capita tonnage estimates unreliable. Finally, while drawing time-series inferences from cross-section evidence is risky,[e] the moderately high income elasticities suggest that continued growth in per capita refuse generation can be expected. Zero population growth cannot be expected to ease the strain on collection, disposal, and recycling facilities.

NOTES

1. The only evidence Gueron is able to obtain is per capita refuse generation for a sample of twelve cities. The city with the lowest per capita refuse generation is also the only one that imposes user charges for collection based on quantity, providing some evidence that "price matters." See Judith M. Gueron, "The Economics of Solid Waste Handling and Government Intervention,," in *Public Prices for Public Products*, ed. Selma Mushkin (Washington, D.C.: The Urban Institute, 1972), p. 184.

2. George R. Davidson, Jr., *A Study of Residential Solid Waste Generated in Low-Income Areas* (Cincinnati: U.S. Environmental Protection Agency, 1972).

3. H. S. Houthakker and Lester D. Taylor, *Consumer Demand in the United States* (2nd ed.; Cambridge: Harvard University Press, 1970).

[e]Such inferences are especially risky here since changes in packaging technology and recycling practices can have a profound effect on tonnage generation.

Chapter Five

User Charges for Refuse Collection

One of the most important decisions facing those cities which provide municipal or contract collection is the choice of the method of financing the collection service. The basic question that must be answered is deceptively simple: How should the costs of publicly provided goods and services be apportioned among the residents of the jurisdiction? In this chapter we address this question with special emphasis on user charge financing. After describing alternative methods of financing and existing financing practices, we analyze the case for user charges in detail.

METHODS OF FINANCING

Although they vary in many particulars, three broad types of financing methods can be distinguished: general revenue, service charge, and user charge.

In this taxonomy, general revenue financing includes several commonly used methods of financing, such as raising funds from *ad valorem* taxes on property, income, sales, and the like; from separate property taxes earmarked for a particular service; or from special assessments.

As an alternative to general revenue financing, some municipalities levy service charges, flat fees which vary with other characteristics of properties, such as the type of land use or the number of dwelling units in a structure. The customary term, service charge, is a misnomer because the charge need not vary with the service provided.

There is an important conceptual distinction between these two methods of financing collection and the third category, user charges. A true user charge varies with the type of service provided to a particular user. The distinguishing test is whether the charge varies with the consumer's choice of service. Implicit in our use of the term user charge is the assumption that the consumer has some choice about those aspects of collection service for which charges vary. For example, if a city chooses to collect residential refuse weekly and commercial refuse daily, and to charge the commercial properties more for their more frequent service, we would call the charge a service charge rather than a user charge, because neither user has a choice of the frequency of collection.

CURRENT FINANCING PRACTICES

All three methods of financing are actually in use. Table 5-1 presents a summary of the available survey evidence on methods of financing contract and municipal refuse collection. In 1964 about half of the jurisdictions surveyed financed collection exclusively out of general revenues. Moreover, although there may be discrepancies between the coverage in the three surveys, there is apparently a trend away

Table 5-1. Methods of Financing Refuse Collection, by Organization, 1956, 1961 and 1964.

	Organization						
	Municipal only		*All other*[a]		*Total*		
Financing Method	*1955*	*1964*	*1955*	*1964*	*1955*	*1961*	*1964*
General tax revenue	71.7%	48.7%	59.2%	51.4%	67.6%	58.6%	50.1%
Service or user charge	28.1	50.8	40.0	47.9	32.3	41.4	49.4
Exclusively		34.7		35.0		29.0	34.9
Partially		16.1		12.9		12.4	14.5
Other	0.2	0.5	0.0	0.7	0.1	0.0	0.6
Sample size	572	429	277	428	849	307	857

[a]Excluding jurisdictions with private collection only, but including jurisdictions with both municipal and contract service.

Sources: Data derived from *Refuse Collection Practices* (3rd ed.; Chicago: American Public Works Association, 1966), p. 271; ibid. (2nd ed.; 1958), p. 200; Lennox L. Moak, *Refuse Collection and Disposal Service Charges* (Municipal Finance Officers' Association, 1961), p. 2.

from the use of general revenues as a means of financing collection. The percentage of jurisdictions with some form of either service- or user-charge financing increased from 32 percent in 1955 to 49 percent in 1964. A similar increase in service- or user-charge financing has occurred in those jurisdictions which provide municipal collection.

From the raw data gathered in the 1961 survey, it is possible to distinguish between jurisdictions imposing service and user charges. Of the 127 jurisdictions which did not finance collection out of general tax revenues, only 25 assessed true user charges against private households. These charges varied with the amount of refuse generated or the type of service chosen, often together with other parameters. Most of the jurisdictions assessing user charges against households were in the western states.

As shown in Table 5–2, general tax financing is the rule in the Northeast. In contrast, in the Southwest and West, general revenue financing is far less common. In the western states only one-sixth of the jurisdictions reported financing refuse collection exclusively out of general revenues.

Financing patterns also vary with city size. Table 5–3 indicates that in both 1961 and 1964 larger cities relied more on general revenue financing than smaller cities.

In short, while all three types of financing are observed in practice, general revenue financing is the most common method used, and true user charges are relatively uncommon. In addition, patterns of financing vary by region and city size. Cities outside the West and Southwest are more likely to finance out of general revenues; larger cities in all regions are also more likely to rely upon general revenue financing.

THE CASE FOR AND AGAINST USER CHARGES

Although user charges are not commonly imposed for refuse collection, economic principles suggest that the imposition of user charges would improve resource allocation in the public sector. By giving consumers the choice of the service they consume and charging them for it, consumers are given an incentive to economize on the resources of collection agencies, improving the mix of collection services that is provided. In contrast, if the government chooses the type of service provided, it has no practical way of evaluating alterna-

Table 5–2. Methods of Financing, by Geographical Region.

Region	General Tax Revenues[a]	Service or User Charge Exclusively	Service or User Charge Partially	Franchise	Total[b]	Sample Size
Northeast	89	6%	6%	0%	100%	62
Southeast	62	23	14	2	100	65
Midwest	64	20	14	1	100	87
Southwest	29	62	6	0	100	45
West	17	39	9	33	100	64

Source: Lennox L. Moak, *Refuse Collection and Disposal Service Charges* (Chicago: Municipal Finance Officers' Association, 1961), Appendices D, E, and F.

[a]Exclusively.

[b]May not add due to rounding.

Table 5-3. Methods of Financing Refuse Collection, by City Size.

A. 1961

	Population			
Financing Method[a]	Over 100,000	50,000 to 100,000	Less than 50,000	Total
General tax revenues	59.4%	64.9%	52.6%	58.6%
Service or user charge	40.6	35.1	47.4	41.4
Exclusively	21.9	25.8	37.8	29.0
Partially	18.7	9.3	9.6	12.4
Sample size	96	97	114	307

[a]For survey respondents with municipal or contract service.

B. 1964

	Population			
Financing Method	Over 100,000	50,000 to 100,000	Less than 50,000	Total
General tax revenues	60.0%	58.0%	47.9%	50.1%
Service or user charge	40.0	40.9	51.6	49.3
Exclusively	25.0	28.0	36.8	34.9
Partially	15.0	12.9	14.8	14.4
Other	0.0	1.1	0.0	0.6
Sample size	80	91	677	850

Source: Data derived from *Refuse Collection Practice* (3rd ed.; American Public Works Association, 1966), p. 271; ibid. (2nd ed.; 1958), p. 299; Lennox L. Moak, *Refuse Collection and Disposal Service Charges* (Municipal Finance Officers' Association, 1961), p. 2

tive mixes of service, and even if it did, it would have very little incentive to do so.

Efficiency gains are not the only argument for user charges. They are supported by some as a way of collecting revenues from properties which are exempt from property taxes, but which would be liable for user charges for the services they consume. This would have the effect of increasing the revenue base or, if these users decided to arrange for private collection, of reducing the total public cost of refuse collection. While this argument is probably not terribly important for residential refuse collection, it may be significant for nonresidential collection services provided by many cities.

While user charges may increase efficiency and raise additional revenue, there are several arguments against them. The administrative costs of calculating, billing, and collecting user charges may be substantially greater than those of financing through a property tax system already in place. These administrative costs might be too great to justify gains in allocative efficiency.

The regressivity of user charges is another argument against them. Depending on refuse generation, charges which vary in proportion to the amount of refuse generated could constitute a much higher proportion of income for the poor than for the rich.

A final argument against user charges results from their federal income tax treatment. Because they are fees rather than taxes, homeowners cannot deduct them from gross income in the calculation of federal income taxes. The imposition of user charges will increase the federal income tax liability of city residents. Thus, federal tax laws create an artificial incentive not to impose user charges.

The weight to be attached to each of these arguments depends on the magnitude of their separate effects. While we cannot estimate the magnitude of the efficiency gains or the administrative costs of imposing user charges, we can present evidence on the magnitude of the cost differentials, the regressivity of the charges, and the magnitude of the tax loss. The studies of the demand for and the cost of collection presented in earlier chapters bear on these questions. Before presenting this empirical evidence, however, we must be precise about specific charges which might be imposed, and their implications for economic efficiency.

EFFICIENCY IN THEORY

The principal argument for user charges is that they will increase the efficiency of resource allocation. If user charges improve efficiency in the provision of local public services, it is because such charges provide incentives for residents to alter their behavior in ways which economize on scarce resources.[a] The efficiency gains will be greatest if each resident is charged the marginal cost incurred in providing the service he consumes. In addition, when marginal cost is below average cost, the total benefits of providing the service to all residents must exceed the total costs; otherwise the service should not be provided at all.

The principle that residents should be charged the marginal cost of

[a]In addition to improving allocative efficiency, by focusing attention on cost, user charges might also be expected to improve the productive efficiency of the collection operation.

the services they consume is complicated by the fact that there are several dimensions to refuse collection services which affect the cost of service provision. In Chapters 2 and 3 we identified three such dimensions: the type of service (i.e., location of pickup), the amount of refuse collected, and the frequency of collection.[b] Applying the marginal cost principle requires that residents be free to alter their consumption in any of these dimensions and that they be charged according to differences in the services they elect. For each dimension of service, the efficiency principle has slightly different implications.

For differences in the type of service, residents who elect back-of-the-house collection should be charged for the additional cost this imposes on the collection agency. Such a system of charges would increase allocative efficiency, since those households which valued back-of-the-house collection at less than its marginal cost would have an incentive to substitute their own labor for that of the collection agency.

For differences in the amount of refuse, the efficiency principle implies that households should be charged the marginal cost of collecting the refuse they generate. This charge provides incentives to reduce refuse generation and to recycle the refuse generated. At the same time, however, this charge provides an incentive for increased littering and accumulation of refuse. Whether such a charge would increase economic efficiency depends upon the tradeoff between the benefit of reduced generation and the cost of additional littering and health hazards.[c]

[b]A fourth dimension of service that affects collection cost is the location of the residence in the city. Because marginal collection cost varies with residential density and distance to the disposal site, tonnage charges should, at least in principle, vary with residential location. The purpose of such a charge would be to provide a greater incentive to reduce generation in those areas where collection is more expensive. It might even be argued that because an individual's residential location decision can affect refuse collection cost, location charges should be imposed to force residents to consider these costs of collection when they make location choices. Such charges would differ from marginal cost charges. In fact, it is difficult to imagine that such charges would have any effect on a household's residential location decision. We have not included a discussion of charges which vary with location because we feel that they are impractical and that the gains from imposing such charges would be small.

[c]This paragraph is subject to a number of qualifications. First, because marginal cost is below average cost, providing collection service must pass the overall benefit-cost test. It seems reasonable to assume that this is the case for once-a-week curbside collection. Second, a tonnage charge would also provide an incentive to substitute garbage disposals for refuse-collection services. Whether

Applying the efficiency principle to the frequency of collection is more complicated. Once a city has decided to offer twice-a-week service, then residents should be charged only the marginal cost of collecting their refuse. But the city must decide whether to provide twice-a-week service according to the overall benefit-cost criterion. In principle, if the demand curve for frequency of collection were known, it would be possible to compute the total benefits and the total cost of twice-a-week as compared to once-a-week service and to provide twice-a-week service if total benefits exceeded total costs. If total benefits exceed total costs, all residents should be charged the same rates per ton of refuse generated. The point is that once the decision to cover the routes twice a week has been made, it is inefficient to charge residents for the more frequent service; they should only be charged the incremental cost of stopping at their house, but this cost is already reflected in the tonnage charge.

In theory, charges which vary along these three dimensions would increase economic efficiency if there were no costs of imposing them. In practice, however, there are substantial implementation and administration costs against which the gains in allocative efficiency must be weighed. In order to reach a conclusion about what charges are practical, the likely magnitude of the implementation costs must be compared to the efficiency benefits of each charge. We postpone an assessment of the practicality of each charge until after we examine the empirical evidence on the other arguments concerning user charges.

EQUITY

The cost of refuse collection is small relative to total income; for the sample of Connecticut towns, the cost of refuse collection is only 0.3 percent of median family income. Thus, the effect of user-charge financing on the distribution of income is hardly the most significant equity issue of general governmental policy. Nonetheless, it may be a serious issue in refuse collection policy, because residents may expect to receive the service "free" and because there is a presumption that general revenue financing, when compared with user-

---

this substitution is desirable depends upon the cost of collection and disposal relative to the cost of sewage treatment and disposal. Third, if littering and the accumulation of refuse could be reduced by increased law enforcement, the appropriate comparison is to the cost of increased enforcement.

charge financing, benefits the poor. A review of the public administration literature indicates that the incidence of user charges is indeed an important consideration in the choice of financing method. For example, in discussing the advantages of general revenue financing, the American Public Works Association (APWA) says that under such a system "the cost of a collection operation is distributed more nearly on an ability to pay basis" that under alternative methods.[1] In discussing the service-charge alternative, the APWA cites the following as the major disadvantage of such charges:

> Refuse service charges levied according to the benefits received are regressive. The amount of refuse does not vary with the ability to pay taxes; poor families may have more refuse than wealthy ones.[2]

No evidence is cited to support the conclusion that user charges are regressive when compared to general revenue financing. While there is no evidence on the incidence of charges which vary with type of service or frequency, we can present some evidence on the incidence of tonnage charges. The analysis presented in Chapter 4 provides strong evidence that lower income households generate less refuse than higher income households. The estimates of the income elasticity of refuse generation can be compared to existing estimates of property tax incidence to indicate the net incidence of user charges.[3]

Table 5–4 presents our estimate of the income elasticity of tonnage charges as a proportion of income together with estimates of the income elasticity of property taxes as a proportion of income derived from existing empirical studies of the incidence of property taxation. As compared with a proportional income tax, where the elasticity is zero, a tonnage charge would be slightly regressive.[d] Property taxes however, are far more regressive. Thus, replacing the property tax with a charge on the amount of refuse collected would be highly progressive.

FEDERAL INCOME TAX DEDUCTIBILITY

The choice of financing method may affect the federal income tax deductibility of payments for refuse collection. If refuse collection is

[d]These estimates hold family size constant. To the extent that the poor have larger families, these estimates would be closer to the property tax elasticities.

Table 5-4. Estimated Progressivity of Financing Refuse Collection through Property Taxes and through User Charges on the Amount of Refuse Collected.

	Estimated Income Elasticity of Charges as a Proportion of Income
User charges	−0.24
Property tax:	
Homeowners	
Netzer	−2.63
Morgan	−1.82
Renters	
Netzer	−1.42
Morgan	−1.68
Combined owners and renters	
Pechman/Okner	−3.05/−4.00

Sources: Computed from Dick Netzer, *Economics of the Property Tax*, (The Brookings Institution, 1966), Tables 3-6 and 3-8; James N. Morgan, et al., *Income and Welfare in the United States* (New York: McGraw-Hill, 1962), Tables 19-1 and 19-6; Joseph A. Pechman and Benjamin A. Okner, *Who Bears the Tax Burden?* (The Brookings Institution, 1974), Table 4-8, Variant 3-b/1c. All estimates were obtained by regression analysis of data over the income range \$0-\$20,000. Table presents the coefficient b estimated from the regression log $(T/Y) = a + b \log Y$, where T is average property tax payment, and Y is annual income. The user charge elasticities were calculated using the New Haven generation equations from Chapter 4.

financed by any uniform *ad valorem* levy, except special assessment levies, the amount is deductible from gross income for all residents. If, however, collection is financed by a service charge that varies with any other characteristics of the property (such as structure type), or by a user charge that varies with the type of service, frequency of collection, or amount of refuse generated, such a charge does not qualify as a deductible personal expense. As a result of these tax laws, a dollar of municipal expenditure on refuse collection financed by taxes costs the homeowner who itemizes deductions less than a dollar. This does not change the real resource cost of refuse collection, but—by transferring some of the tax burden to the federal level—it makes property tax financing more attractive from the city resident's perspective than either service charges or user charges.

How important is this federal tax incentive for financing out of general revenues? The magnitude of the saving available to any individual depends upon his marginal tax bracket and on whether he itemizes his deductions. Table 5–5, which presents the marginal tax brackets associated with income classes, indicates that a community

Table 5-5. Estimated Tax Saving Due to Federal Deductibility of Local Taxes

		New Haven			Woodbridge		
Income Class (thousands of dollars)	Marginal Tax Rate[a]	Number of Households[b]	Estimated Proportion Itemizing[c]	Rate of Tax Saving	Number of Households[b]	Estimated Proportion Itemizing[c]	Rate of Tax Saving
0- 3.9	0.14	21,933	12%	02%	71	49%	07%
4- 5.9	0.16	6,998	19	03	58	60	10
6- 9.9	0.19	12,295	26	05	149	72	14
10- 14.9	0.22	9,690	40	09	345	83	18
15- 19.9	0.25	4,931	55	14	617	89	22
20- 24.9	0.28		55	16		89	25
25- 29.9	0.32	1,338	66	21	548	95	30
30- 49.9	0.39		66	26		95	37
50 99.9	0.53	283	66	35	161	95	50
100-199.9	0.62		66	41		95	59
200+	0.70		66	46		95	66
Tax on residential properties: average tax savings to households				06			27
Owner-occupied property as a proportion of all taxable property[d]				15			80
Tax on all properties: average tax savings to households				01			21

[a]U.S. Department of the Treasury, Internal Revenue Service, *Statistics of Income 1971*, "Individual Income Tax Returns," Publication 79 (Washington, D.C.: U.S. Government Printing Office, 1973), p. 350.

[b]U.S. Bureau of the Census, *1970 Census of Housing*, "General Housing Characteristics," Final Report HC(1) - A8, *Connecticut.* (Washington, D.C.: U.S. Government Printing Office, 1971), Tables 10 & 24.

[c]For New Haven, the estimated proportion of households itemizing deductions is the proportion of homeowners, by income class found in U.S. Bureau of the Census, *1970 Metropolitan Housing Characteristics* (Washington, D.C.: U.S. Government Printing Office, 1972), Tables A-3, B-3, C-3. For Woodbridge, the proportion used is that for metropolitan area residents living outside New Haven and West Haven.

[d]State of Connecticut, *Information Relative to the Assessment and Collection of Taxes*, Public Document No. 48 (Hartford: State of Connecticut, 1970).

composed entirely of households in the \$20,000 to \$25,000 income bracket who itemize deductions would have a tax savings of 28¢ per dollar of property tax paid.

More generally, the percentage tax savings available from general revenue financing in any jurisdiction depends upon the income distribution and the proportion of residents who itemize deductions. Table 5–5 indicates the income distribution of New Haven (with a median household income of \$5,943 including families and unrelated individuals) and that of the high-income bedroom suburb of Woodbridge (median income \$17,956). Also presented are estimates of the proportion of homeowners by income class, a proxy for the fraction itemizing deductions. As the table indicates, not only are household incomes and marginal tax brackets higher in Woodbridge, but home ownership levels (and hence the proportion who itemize deductions) are also higher.

If we consider a service provided only to residential properties financed by a tax on those properties, the weighed average of the tax savings from financing through general revenues is 0.27 in Woodbridge, indicating that local residents would pay only 73 percent of the costs of service provision. The corresponding figure for New Haven indicates that residents would pay 94 percent of the costs of this service.

On the other hand, if we consider an identical service financed by a general tax on all properties, the average tax saving to residents also depends upon the proportion of taxable property devoted to rental units and business enterprise, since either kind of charge is deductible as an ordinary business expense for these types of properties. In New Haven, if a given service provided to residences, were financed out of general revenues instead of by user charges, the dollar cost to local residents and businesses would be reduced by about 1 percent. In the nearby suburb, the corresponding tax saving would be more than 20 percent.

In short, although it is not clear whether people fully perceive it, a substantial tax incentive exists to finance refuse collection — or any public service — through property taxes or other *ad valorem* taxes. For richer communities, the incentive appears to be quite strong.

REVENUE FROM TAX-EXEMPT INSTITUTIONS

One motivation for imposing user charges is to raise revenue from tax-exempt institutions. One town in Connecticut has imposed service charges with precisely this motivation. The city of Middletown organized a special district for the purpose of refuse removal under specific state statutory authority. This special district makes charges to properties of various classes (e.g., churches and schools). The special district also offers a choice of 2, 5, or 6 collections a week to commercial establishments. The charge for service varies with the square footage of the structure. All commercial establishments must pay the established rate for at least twice-weekly collection even if they do not use the service. In addition, the special district sets service charges for twice-weekly residential collection. These charges vary with the number of dwelling units on any parcel. Figure 5–1 shows how charges vary with frequency of collection, structure type, and land use.

These charges are service charges with respect to the amount of refuse generated but are true user charges with respect to the frequency of commercial collection (except for the required minimum payment for twice-weekly collection). For the purpose of raising revenue from tax-exempt property, however, there is no difference between service and user charges.

City officials report that 3.9 percent of the total revenue collected through these charges comes from tax-exempt property, indicating that a modest amount of additional revenue can be gained from tax-exempt institutions. Although economists are not inclined to attach much weight to arguments based solely on revenue considerations, city officials facing budget deficits clearly do. The Middletown story lends some credence to this argument for user charges.

EFFICIENCY IN PRACTICE

In theory, user charges would lead to a more efficient allocation of resources. In practice, however, the cost of imposing the charges may be so great that they outweigh the gains in allocative efficiency. Thus some analysis of the likely magnitude of the administrative costs and the gains in allocative efficiency is required. The gains in allocative efficiency depend upon the responsiveness of demand to changes in price and upon the magnitude of the cost difference. Each of these factors is discussed for the three types of charges.

Type of Service

A user charge for back-of-the-house service could probably be imposed with relatively modest additional administrative expense.

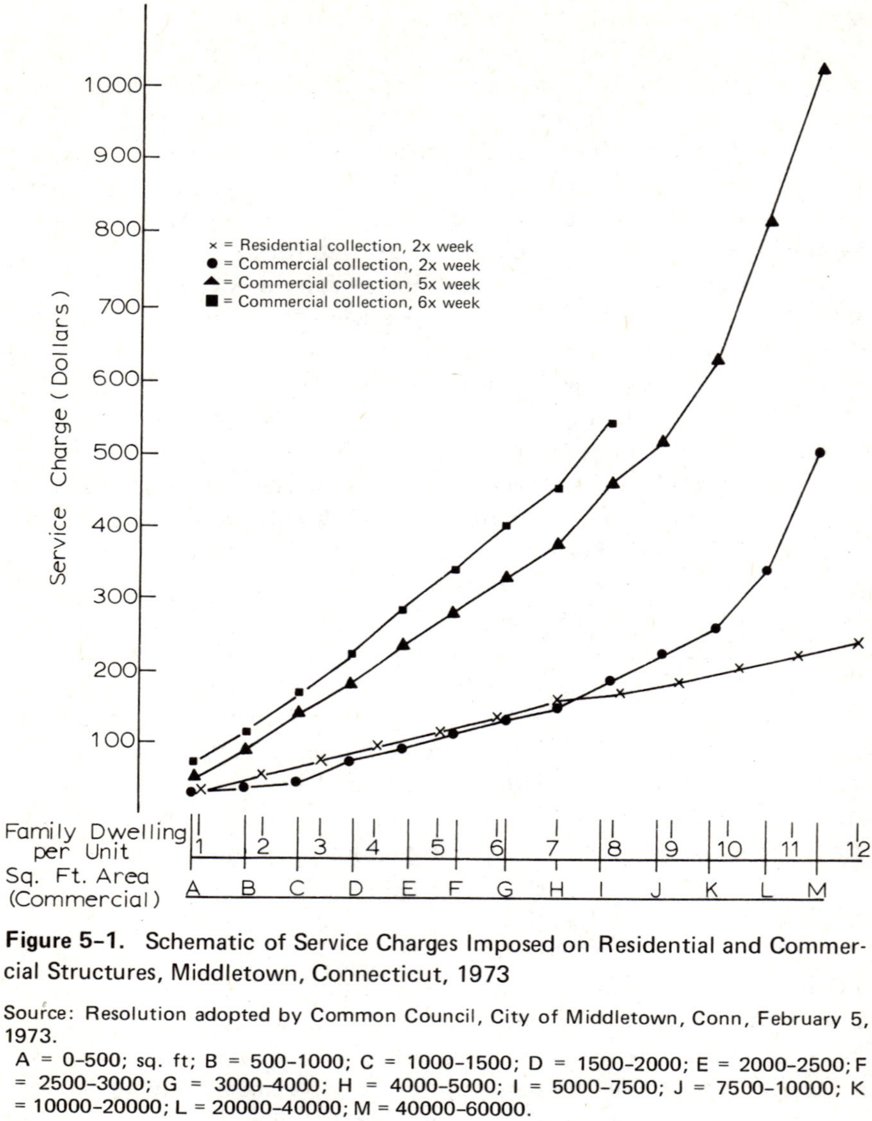

Figure 5–1. Schematic of Service Charges Imposed on Residential and Commercial Structures, Middletown, Connecticut, 1973

Source: Resolution adopted by Common Council, City of Middletown, Conn, February 5, 1973.
A = 0–500; sq. ft; B = 500–1000; C = 1000–1500; D = 1500–2000; E = 2000–2500; F = 2500–3000; G = 3000–4000; H = 4000–5000; I = 5000–7500; J = 7500–10000; K = 10000–20000; L = 20000–40000; M = 40000–60000.

Crews would need to maintain a record of those households which had chosen back-of-the-house service, but this is certainly feasible. At least one jurisdiction in Connecticut already provides two levels of service; residents over 65 years of age are provided back-of-the-house collection and others receive curbside service. In addition, households would have to be billed for the more expensive service. One simple way of doing so would be to use distinctive markings on refuse containers or color-coded plastic bags. Municipalities could sell plastic bags color-coded for back-of-the-house collection at a higher price reflecting the additional cost of this service.

The gains in allocative efficiency from imposing charges for back-of-the-house service are likely to be substantial. On the demand side, we suspect that the choice of service is quite responsive to changes in price. On the cost side, the difference in cost between back-of-the-house and curbside service is quite large. The intercity regressions indicate that back-of-the-house service is between 33 and 47 percent more expensive than curbside service, after controlling for other factors. These two facts together imply that the efficiency gains are likely to be relatively large.

In short, the small administrative cost and the large gains in allocative efficiency make a strong case for charges which vary with the type of service.

Tonnage

Any difficulties in implementing charges which vary with the amount of refuse collected are also likely to be relatively small. Again, specially marked plastic bags could be sold at a price representing the marginal cost of collection. Alternatively, drivers could keep tallies on prepared lists noting the number of 30-gallon cans which were emptied at each residence, to be used later for direct billing. It should be noted that some cities already do impose charges which vary with the number of cans.

The efficiency gains from tonnage charges are much more difficult to assess. The amount of refuse generated may not be very responsive to changes in price. Furthermore, some residents may simply be unwilling to pay to have refuse collected. They may instead dispose of their refuse by dumping it in parks, abandoned lots, or their own yards. The resulting social costs can be quite high, especially in high-density neighborhoods.

The evidence on cost is much clearer. Although the marginal cost of collecting is below average cost, it is nonetheless quite high. For the cities of New Haven and Hartford, marginal cost is estimated to be roughly 56 and 85 percent of average cost, respectively. In addition to these collection costs, an amount to cover the cost of disposal should be added, further increasing the charge.

In conclusion, while the administrative costs are likely to be relatively small and the marginal cost of collecting additional tonnage is substantial, the magnitude of gains in allocative efficiency cannot be assessed without direct estimates of the effect of charging on the amount of refuse generated and the amount of littering done. Thus there is considerable risk of reducing efficiency by imposing tonnage charges, especially in high-density areas where accumulation of refuse is a more serious problem.

An alternative tonnage charge levied on producers rather than consumers has much greater promise for reducing refuse generation, and it would not create an incentive to litter. The federal government could impose a charge to cover the cost of collecting and disposing of a product when it is finally discarded. In principle, the charge could vary with the composition of the product and the estimated timing of its ultimate disposal. We suspect that a charge nominally levied against the producer would have a considerable effect on the amount of refuse generated. In short, a federal "waste tax" levied on producers appears to merit serious consideration.

Frequency

As noted above, there is no efficiency argument for imposing a charge for twice-a-week collection (other than the tonnage charge). The important decision for efficiency is whether to offer more frequent collection. For this decision the efficiency criterion is that the total benefits of more frequent collection exceed the total costs. In practice it would be difficult to conduct such an analysis rigorously at the local level. This leads us to search for some approximation to the benefit-cost calculation that can be institutionalized in a simple manner. Average cost pricing of twice-a-week service is a conservative strategy that does force households to reveal their preferences. If, as a group, households choosing twice-a-week collection pay the added cost of covering the routes a second day each week, then the total benefits necessarily exceed the total cost. Although this is clearly not a fully efficient charge, it does provide a mechanism for ensuring that more frequent service is not offered unless it is justified.

The difficulty of implementing such a charge would not be great. Drivers would have to keep track of which houses are to receive more frequent collection, and there would be some additional billing cost, but it does not appear to be substantial. Moreover, as noted above, charges which vary with the frequency of collection are already in use in some towns.

The gains in allocative efficiency are difficult to assess. On the cost side, the intercity cost analysis indicates that there is a substantial difference in total cost between once- and twice-a-week collection. Depending upon which measure is used, the difference in total cost ranges from 34 to 88 percent of once-a-week service and averages roughly 50 percent.

On the demand side, the choice of twice-a-week collection is probably relatively sensitive to differences in price and income. If so, benefits will vary not only across cities but also within cities. This implies that in large cities any benefit-cost calculation should be undertaken separately for routes or neighborhoods rather than for the city as a whole. Since neighborhoods tend to be segregated by income and tastes, allowing different collection frequencies across neighborhoods could substantially increase efficiency. Operationally, the efficiency gains could be achieved by establishing refuse collection districts with charges varying by frequency of collection. More frequent collection would be offered in any district where the total revenues from average cost pricing of more frequent service exceeded total cost. In larger cities, therefore, experimentation with smaller districts with different collection frequencies and correspondingly different charges has potentially large benefits.

In summary, we propose charging households the average cost of providing more frequent pickups as a second-best method of insuring that the benefit-cost criterion for efficiency is met. The administrative costs do not appear large, the cost of collection does vary significantly with frequency, and the demand for more frequent collection probably varies substantially with price. Whether, in practice, average cost pricing would perform better than the political process in insuring that the benefit-cost criterion is met, is more difficult to assess.

CONCLUSION

We conclude that there is a strong case for the imposition of user charges for refuse collection, at least for charges which vary along certain dimensions. In particular, it appears that user charges which vary with the type of service and the frequency of collection are likely to result in an improved allocation of resources. Charges which vary with the amount of refuse generated by households could also lead to improved efficiency, but because they create an incentive for increased littering, they should probably not be imposed in high-density areas. Federal charges which vary with the amount of the product that will ultimately be disposed of as waste should be considered as an alternative to local tonnage charges.

Although the evidence supporting these conclusions is not definitive, the analysis does indicate that the cost differentials associated with the type of service and the frequency of collection are substantial. Moreover, we suspect that consumers' choices are sensitive to differentials in the prices of these aspects of collection service. It also appears that such a system of charges is administratively feasible.

One strong argument against user charges, however, is that unlike property taxes, they do not qualify as personal deductions for federal income tax purposes. Consequently, from the perspective of city residents, user charges are more costly than property taxes, especially for high-income residents. Given the potential efficiency gains, this provision of the Internal Revenue Code merits review for possible elimination of this artificial incentive.

NOTES

1. American Public Works Association, *Refuse Collection Practices* (3rd. ed.; Chicago: Public Administration services, 1966), p. 271

2. Ibid., p. 244

3. The incidence of property taxes by income class is not without some ambiguity. This ambiguity arises because real property taxes are partly a captial tax, the incidence of which is unknown, and because the extent of capitalization into land values is unknown. For a discussion of these issues, see Peter Mieszkowski, "The Property Tax: An Excise Tax or a Profits Tax?" *Journal of Public Economics* 1 (April 1972): 73-90.

Chapter Six

Empirical Findings and Evaluation[a]

The research reported in this volume summarizes the available evidence on the delivery of residential refuse collection in urban areas and, using data specially collected from jurisdictions in Connecticut, presents a more detailed economic analysis of the organization and financing of collection activities. This chapter summarizes the major empirical results on the economics of refuse collection, reported in detail in the previous chapter, and indicates some of their implications. The chapter concludes by noting several areas where our knowledge could be improved.

SUMMARY OF EMPIRICAL FINDINGS

The principal empirical findings of this research fall into three categories: (1) existing practices; (2) costs; and (3) demands.

Existing Practices

- As currently practiced, refuse collection is an extremely labor-intensive process. On average about 73 percent of total resource cost is accounted for by the wages of crews and their direct supervisors.
- Municipal budget and expense data significantly understate the actual costs of collection. True resource costs are typically 20 to 40 percent higher than budget figures due to the omission of such items as fringe benefits, the cost of capital goods, and administrative costs.

[a]After receiving the proofs of this monograph, we became aware of research by E.S. Savas of Columbia University which apparently corroborates many of our findings and evaluations. Although we have not had access to Savas' work, and cannot comment on the evidence supporting his conclusions, his research findings are apparently similar to our own.

· Appropriation from general revenue is the most common method of financing collection activities, especially in the Northeast, but its use appears to be declining over time. True user charges are, however, relatively rare.

· The deductibility of property taxes from gross income under the Internal Revenue Code provides a surprisingly strong incentive for financing refuse collection, or any other service, out of general revenues. The strength of the incentive varies with the income distribution of residents and the proportion who itemize personal deductions. In the central city of New Haven, for example, tax deductibility reduces the cost to city residents of tax-financed public services by about 1 percent, but in the affluent suburb of Woodbridge, the cost is reduced by about 20 percent.

· Exemptions of municipal collection agencies from taxation provide a significant incentive for municipal provision of service. Tax exemptions make municipally provided service appear to be about 6 percent cheaper from the perspective of city residents than private service using the same resources.

· The existing organization of collection varies systematically with the characteristics of jurisdictions. As city size and density increase, the most prevalent form of organization changes from private to contract to municipally-provided service.

Costs

· Collection cost varies markedly with the type of service provided and the frequency of collection. Holding other factors constant, the average cost of back-of-the-house collection is approximately 40 percent higher than curbside collection. Twice-a-week collection is roughly 50 percent more expensive than once-a-week service.

· Collection cost appears to vary systematically with the organization of collection, even after controlling for other factors. Private collection appears to be about 30 percent more expensive than municipal collection, which, in turn, appears to be about 25 percent more expensive than contract collection.

· Refuse collection appears to be characterized by economies of density. A doubling of the tonnage collected per mile of collection route may reduce average cost by up to 50 percent.

· There is no evidence of significant economies of scale in refuse collection.

· Longer hauls to the disposal site apparently increase average cost significantly, but consistent estimates of the magnitude of the increase could not be obtained using the available data.

Demand

- The quantity of refuse generated varies markedly with the incomes and family sizes of residents. A doubling of family income leads to an increase of roughly 50 percent in refuse generation.
- User charges for the amount of refuse collected are likely to be mildly regressive relative to income, but substantially less regressive than financing by property taxes.

IMPLICATIONS

The empirical results highlighted above are subject to many qualifications noted in previous chapters and to verification by future research. Nevertheless, they provide the basis for several conclusions about directions for policy initiatives at the local level and at higher levels of government.

Organization

Contracting for refuse collection services appears to be cheaper than either municipal or private collection, even after controlling for density, service, and frequency of collection. These results suggest that municipalities should seriously consider contracting with private firms for refuse collection services. Careful safeguards need to be established to ensure that bidding for contracts is in fact competitive and that barriers to entry by new firms are minimal. Cities with small populations which presently permit private collection with little or no regulation should consider competitive bidding to reap any economies of density and to enforce competitive behavior. Cities with large populations might consider dividing the city into several sectors for which separate bids would be solicited on a staggered basis, thereby keeping entry barriers modest.

Management Information

Managers of city collection agencies are often forced to operate without any objective standard of performance. Not only are costs imperfectly estimated, but tonnage and volume collected are often not recorded. Without knowledge of the output produced or the cost of operation, any inefficiency that exists can scarcely be blamed on agency managers. Moreover, even if more detailed information were available, a yardstick against which to measure performance is missing.

A first step in providing management with needed information could be taken at little expense by the federal government. A modest investment could produce consistent cost and output reporting systems which would provide each city with useful performance measures as well as standardized data on costs across jurisdictions. Public tabulations of average cost for cities with different population densities, types of service, and frequencies of collection would provide managers with some comparative data for the evaluation of agency performance. A more ambitious program would provide direct assistance to large cities to develop management information systems, which would improve the analysis of crew performance and the search for increased productivity. Pilot experiments in both of these areas would be relatively inexpensive.

Financing

There appear to be advantages to financing refuse collection by specific charges which vary with the type of service provided and allowing residents to choose the type of service they receive. According to the variations in the cost of collection and the demand for service, these charges should vary with location of pickup (i.e., curbside versus back-of-the-house), frequency of collection, and perhaps with the quantity of refuse collected. Experimentation with user charges for refuse collection appears to be warranted, especially since the administrative costs of user charges appear to be modest.

The deductibility of "taxes" and the nondeductibility of "charges" for federal income tax purposes provides a strong, artificial incentive to maintain property tax financing, even in the face of efficiency gains through user-charge financing. Serious consideration thus should be given at the federal level to treating "taxes" and "charges" for the same service on an equal footing, perhaps by eliminating the deduction for taxes.

Planning for Future Growth

The amount of refuse generated is sensitive to levels of income. Therefore officials concerned with future collection and disposal practices should assume that the amount of refuse generated will continue to increase more rapidly than population.

Efforts to reduce the rate of increase in the amount of refuse generated might include the imposition of taxes by the federal government to discourage production of goods with a high waste content, notably paper products and packaging.

Disposal Practices

Disposal and collection are highly interdependent. The impact of proposed changes in disposal practices upon collection costs can be substantial. For example, changing the location of the disposal site, by increasing hauling distances, will increase collection costs. Similarly, the separate collection of newspapers, glass, or metal for recyling, by reducing effective density, will increase collection cost. These cost differentials are nontrivial and cannot be ignored in making decisions on disposal practices.

UNRESOLVED ISSUES

It is worth noting in conclusion some of the limitations of the analysis and several issues which are worthy of future investigation.

The cost analysis presented in Chapter 2 is based upon a simple general model. Although the empirical results appear to be useful and promising, it is clear that more sophisticated analysis, undertaken in conjunction with engineers with practical experience, could increase our understanding of the way costs are incurred and improve our estimates of cost functions.

Better data should be collected. In many places in the preceding chapters it was necessary to make qualified inferences from data that were not adequate to address an issue conclusively. In some cases, simply increasing the sample size could have produced more conclusive results. For example, a larger sample would presumably have permitted us to separate with greater confidence the effect of density on collection cost from that of organization.[b] In other cases, however, additional information is needed. For example, information on the wage rates paid by private firms might provide additional insight into the reasons for cost differences by type of organization.

In conclusion, there are several important issues on which we were

[b]In fact, we devoted a great deal of effort to improving the quality of the data by reducing possible errors of measurement, especially in the intercity samples used in the cost and demand analyses. In retrospect, our resources would have been better spent increasing the sample size by surveying more cities from different states.

not able to present any evidence, which also merit further investigation. An assessment of alternative technologies, such as containerized collection, should be undertaken. A study of the sources of invention and the spread of new technologies throughout the sector deserves high priority. Labor relations, a glaring omission in a sector that is so labor intensive, should be examined, with emphasis on the effect on unionization on wage rates and productivity. Finally, the analysis should be extended to nonresidential refuse collection.

Appendix A

Data Collection for the Intracity Analysis

Parts of the intracity analyses discussed in Chapters 2 and 4 are based on data derived from internal records maintained by the Public Works Departments of two Connecticut cities, New Haven and Hartford. This appendix briefly notes the derivation and definitions of the variables used in the main text from the raw data.

THE HARTFORD DATA

The Department of Public Works keeps detailed daily time cards on its operations. In particular, each truckdriver records the time he begins collecting each load of refuse, the time he finishes collecting, and the time each load is dumped at the incinerator. The weight of each load is recorded, together with the truck identification number and the names of crewmen. Figure A–1 is a facsimile of the records maintained for each collection truck for each day of operation. From these time cards it was possible to compute for each truck the time spent collecting refuse, the time spent hauling, the number of trips to the disposal site, and the amount of refuse collected.

For collection operations, the city is organized into three sectors, each of which is divided into five divisions. One division of each sector is served on each workday. Thus, there are 15 routes, which cover the entire city and receive once-a-week collection. Typically, five trucks service each route; from the records maintained by each driver it was possible to map each vehicle to the route it covered on each day.

For each of the three routes covered on each workday, five vari-

Department of Public Works

Report of Motor Trucks on Ash and Rubbish Collection

Truck No. ____________ Route ____________ Date __________ 19 ___

	Time Began Loading	Time Finished Loading	Time Weighed	Net Weight of Load	Time Returned to Job
1st load					
2nd load					
3rd load					
4th load					
5th load					
6th load					
7th load					
8th load					
9th load					
10th load					

cd197

These columns to be used by by weigher only.

Mileage in ____________ Gasoline ____________ Gals.

Mileage out ____________

Total ____________

Loaders ____________

Remarks: ____________

Signature of Driver ____________

Figure A–1. Records Maintained for Each Refuse Collection Vehicle for Each Day of Operation in Hartford

ables were computed from the drivers' records. These variables, which are used in the cost analysis in Chapter 2, are: total tonnage collected, Q; total time spent collecting, T; pickup time, T_p; haul time, T_h; and the number of loads, N.

Total tonnage collected was computed simply by aggregating the "net weight of load" for each load for each of the five trucks serving each route.

Total time spent collecting was computed by subtracting the "time began loading" for the first load from the "time weighed" for the last load for each truck and summing across the trucks serving each route.

Pickup time was computed by subtracting the "time began loading" for each load from the "time finished loading" for the same load. These times were then aggregated for each load and for all trucks serving each route.

Haul time was computed by subtracting the "time finished loading" for each load from the "time began loading" for the next load. For the last load for each truck, haul time was computed by subtracting the "time finished loading" from the "time weighed." These times were then aggregated for all trucks serving each route.

The number of loads refers to the total number of loads dumped by all trucks serving each route.

When aggregating from the raw data for each run for each truck there were frequently one or more missing items. When a single haul time was not recorded on the driver's record, it was imputed as the average of haul times for that truck for that day. When a single pickup time was not recorded, it was imputed as the average pickup time per ton for other runs of the same truck. When two or more data items were missing, it was often necessary to delete the observation on that route-day from the sample.

In addition it was necessary to delete observations on routes served by fewer than five trucks from the sample. Fewer than five trucks was indicated every time a single time card was missing from the Hartford records as well as when fewer than five trucks actually served a route on any day.

THE NEW HAVEN DATA

In New Haven, the Department of Public Works maintains similar, but less detailed, daily records on its operations. For each truck a record is maintained indicating the weight of each load and the time it was dumped at the incinerator. Figure A–2 is a facsimile of the records maintained at the New Haven incinerator.

For each of the 11 routes served on any workday, three variables were computed from these records: total tonnage collected, Q; total time spent collecting, T; and the number of loads, N. The tonnage figures are used in the analysis presented in Chapter 4; all three variables are used in the analysis in Chapter 2.

In comparison with the data analyzed for Hartford, the information available for New Haven did not require aggregation over individual trucks and thus required no imputations when parts of the raw data were missing. The raw data does not indicate pickup time and haul time separately, however. In addition, the total time spent in refuse collection activities is not known with precision, since the exact time that collection operations began each day is not recorded.

To estimate total time, it was necessary to make an assumption about the time refuse collection operations begin in the morning. Public works officials in New Haven indicated that collection begins between 4:00 and 5:00 a.m., but that there is a certain amount of variation in the beginning times. However, as Figure A–2 indicates, the time at which the first and all subsequent loads were dumped is available for each truck for each day. From this information the starting time was estimated by subtracting the average elapsed time between the first and second dumps (137 minutes) from the average time the first load is dumped (6:52). The assumed starting time for each truck is thus 4:35 a.m.

Since the actual starting times vary somewhat, this assumption introduces a certain amount of measurement error into the analysis in Chapter 2. The principal effect of this measurement error is to increase the variance of the error term in the statistical models, to reduce the goodness of fit of the cost function, and to bias the significance tests of the regression coefficients downward. There is little reason, however, to expect the measurement error to vary systematically with the independent variables used in Chapter 2, and thus to bias their coefficient estimates.[a]

[a]A possible exception may be the variable measuring snowfall. Crews may begin collection operations earlier if snow is anticipated, especially if snow makes refuse collection more time-consuming. New Haven officials suggested that this may be the case when heavy snow has already begun to accumulate in the evening.

Time & Tonnage Refuse Division Week Ending ________ 19 ____

	Monday		Tuesday		Wednesday		Thursday		Friday		Saturday		Total	
	Time	Weight	Time	Weight	Time	Weight	Time	Weight	Time	Weight	Time	Weight	Time	Weight
Route No.														
Driver														
Truck No.														

Figure A-2. Facsimile of Detailed Records Maintained by New Haven for Each Truck

Appendix B

Estimating the Costs of Municipally Provided Refuse Collection Services: Methodology and Cost Comparisons

The estimation of the cost functions presented in Chapter 2 and the analysis of organizational and financing questions in Chapters 3 and 5 required estimates of the economic costs of operating municipal refuse collection services for a sample of cities and towns. These estimates had to be reasonably accurate and based on comparable accounting methods in order to make meaningful comparisons among the costs of municipally provided service, contracted service, and privately provided collection service.

It was immediately apparent that the budgets published annually by municipalities themselves would be inadequate for this purpose. To be sure, when a municipality operates a refuse collection service, there is generally a separate figure for refuse collection in the annual budget. Unfortunately, however, these budget figures typically do not include a number of significant items of cost that are appropriately attributed to the provision of the service. Furthermore, the particular choice of items to be included in the budget figure for refuse collection varies from one municipality to the next.

Perhaps the primary reason that a number of significant cost items do not appear in the municipal total for refuse collection is that there is a distinct limit to the extent to which municipalities pursue "program budgeting" techniques. That is, municipal accounts often group expenditures according to the type of item for which the expenditure was made, rather than according to the service whose operations gave rise to the expenditure. For example, payments into the state workmen's compensation fund typically appear only as a single figure in the municipal budget; seldom are they allocated to the various operating services. Similarly, expenditures for motor vehicle maintenance frequently appear only under a single separate heading for public works rather than being allocated to the various services that operate the vehicles needing repair.

In addition, refuse collection budgets often include additional items—for example the costs associated with the disposal of solid wastes—which must be excluded to obtain an estimate of collection costs alone.

A second reason that municipal budgets frequently understate the cost of refuse collection is that cities seldom engage in systematic capital budgeting. Thus, the cost of capital equipment is rarely amortized over its useful life; instead, the expenditure is simply attributed, in its entirety, to the year of purchase.[a] The result can be significant variations in the recorded cost of a service from year to year, according to the pattern of capital expenditures.[b] Furthermore, municipalities virtually never attempt to estimate and attribute to their separate operating services the opportunity cost or interest cost of their investment in capital equipment.

Because of these difficulties, we found it necessary to make an independent assessment of the cost of refuse collection for a sample of towns which operate such a service. We proceeded in three stages. First, we undertook a fairly detailed study of the cost of refuse collection in a single city, Hartford, Connecticut, in order to (1) gain some perspective on the precise difficulties involved in estimating costs; (2) to establish yardsticks for estimating costs in other municipalities; and (3) to obtain particularly accurate cost information for use in the intracity analysis of collection costs reported in Chapter 2. Second, we undertook a more limited study of collection costs in the town of North Haven, in order to broaden our perspective on the problems involved and, more importantly, to get a better idea of the kind of information we might reasonably ask municipal officials to provide us in response to a questionnaire. Third, we drew upon our experience in Hartford and North Haven to design a questionnaire to be mailed to each of the other twenty-five municipalities in Connecticut that operates its own refuse collection service. The

[a]One exception occurs when the expenditure is financed by a special bond issue.

[b]Sometimes, in addition, capital expenditures are not accounted for according to service at all, but rather simply lumped together in a separate municipal account for "capital expenditures," thus creating a problem of program budgeting as well as capital budgeting.

responses from these questionnaires were then used as the basis for preparing an independent estimate of costs for each town.

These efforts are described in more detail in the pages that follow. The first section describes the methods by which cost estimates were prepared for the town of Hartford. The second section then discusses the approach used in developing cost estimates through the use of the questionnaire. It presents a summary of the information obtained from the questionnaires, including a summary of our estimate of collection costs compared with the figures published by the towns themselves. The final section presents estimates of the tax saving for Hartford due to the exemption of municipal operations from federal and state taxes.

THE COST OF REFUSE COLLECTION IN HARTFORD

The city of Hartford collects refuse from most residences within the city, from buildings owned by the municipality, and from some commercial enterprises. With the cooperation and assistance of several local officials, basic data covering the entire refuse collection operation were obtained form the 1972-73 fiscal year. The cost estimates derived from that information are presented in Table B-1. The derivation of each of these figures is discussed item by item in the following pages.

As shown in Table B-1, our best estimate of the total cost of refuse collection in Hartford for the fiscal year 1972-73 is $1,672,000. The figure published by the city as the total expenditure for "refuse collection" for the same period is $1,185,855. The city, of course, is well aware that its published figure does not represent a complete allocation of costs. The comparison is made here simply to estimate the magnitude of the costs that are in fact incurred in operating the city's refuse collection service but are reflected elsewhere in the city's accounts, or—in the case of interest costs—are omitted entirely.

Direct Labor

The city's budget includes as a separate item within its accounts for refuse collection the total actual expenditure for wages[c] during

[c]This includes what the city terms "longevity payments," which are a form of bonus.

Table B-1. Estimated Cost of Refuse Collection in Hartford, Connecticut, Fiscal Year 1972-73.

Category	Budget Figure	Estimated Cost		
		Lower Bound	*Best Estimate*	*Upper Bound*
1. Direct labor				
a. Salaries	$1,111,855	$1,000,628	$1,000,628	$1,000,628
b. Fringes & uniforms	0	259,157	259,157	259,157
2. Administration	12,130			
a. Operating services, salary & fringes	0	9,435	9,435	9,435
b. Other, salary & fringes	0	14,612	14,612	26,902
c. Overhead	0	1,461	4,822	17,755
3. Capital cost of vehicles	61,870[a]			
a. Depreciation	0	47,754	47,754	73,930
b. Interest cost	0	12,216	24,431	26,925
4. Vehicle maintenance & operation				
a. Direct labor & materials	0	119,106	178,659	248,138
b. Fringe benefits	0	17,351	26,027	36,149
c. Administration	0	26,262	39,393	54,713
d. Insurance	0	10,138	10,138	10,138
5. Buildings, land, & equipment				
a. Depreciation	0	9,918	21,160	33,062
b. Interest cost—building	0	13,093	30,464	51,578
c. Interest cost—land	0	1,818	4,848	9,090
d. Insurance	0	394	526	658
Total Estimate	$1,185,855	$1,543,393	$1,672,054	$1,858,258

[a]Purchase cost of vehicles.

the 1972-73 fiscal year for all personnel working full time on refuse collection. The total came to $1,000,628. This figure includes wages for sanitation laborers, truck drivers, foremen, and a sanitation supervisor and his assistant. The city must also bear, however, the cost of fringe benefits, including Social Security, retirement pay, life insurance, medical insurance, and workmen's compensation. The cost of these items is not included in the budget total for refuse collection. The city estimates, however, that these benefits cost, on the average, an additional 25 percent of wages.[d] This would come to $250,157 for refuse collection workers for the 1972-73 fiscal year. Another item that does not appear in the refuse collection budget is the cost of uniforms for the laborers, which cost the city approximately $9000 during the 1972-73 fiscal year.

The sum of these three items—wages, fringes, and uniforms—yields a total expenditure of $1,259,785 for full-time direct labor.

Administration

The figures presented in the preceding section include the cost of all supervisory personnel who work full time on refuse collection. There are, however, administrative and clerical personnel on two levels within the Public Works Department who devote part of their time to refuse collection and whose compensation is not included in the refuse collection budget.

Operating Services Administration. The Operating Services Administration consists of three employees who handle general administration both for refuse collection and for three other operating services within the Public Works Department. Their wages, reported elsewhere in the municipal budget for 1972-73, were $29,560. Adding 25 percent for fringe benefits brings this to $36,950. It was assumed that these three individuals allocated their

[d]The city breaks down the 25% figure as follows:

FICA	5.85%
Retirement	14.00
Life and Medical Insurance	5.00
Workmen's Compensation	0.20
Total	25.05%

Since the cost to the city of these fringe benefits is not dependent only upon the salary a worker receives, these percentage figures are of course only an approximation.

time among the different operating services in proportion to the size of the budgets for those services.[e] Since 21 percent of the total expenditures for those services was devoted to refuse collection, 21 percent, or $7,760, of the wages and fringes paid for Operating Services Administration was assigned to refuse collection.

To this figure was added a fraction of the wages and fringes paid for the public works dispatcher ($1,674),[f] bringing total wages and fringes for Operating Services Administration that are attributable to refuse collection to $9,435.

Other, Salary and Fringes. There are twelve individuals engaged in administrative and clerical work for the Public Works Department as a whole, which includes not only the operating services but also engineering services and public buildings. Wages for these individuals are included in a separate account for Public Works Department administration, and, as with the Operating Services Administration, are not allocated to a function in any of the city's budget figures. Consequently, we imputed their costs in a manner similar to that noted above.

The 1972-73 total wages and fringes (25 percent) for these individuals was $180,431. Multiplying this figure by the fraction of the public works budget (excluding administration) taken up by the operating services (71 percent) and then again by the fraction of the operating services budget (excluding administration) taken up by refuse collection (21 percent) gives an expenditure of $26,902 that might reasonably be allocated to refuse collection.

However, Public Works Department personnel suggested that only five of the twelve individuals in the departmental administrative unit spent more than a negligible amount of time on refuse collection,[g]

[e]An alternative estimating procedure based on a slightly different allocation of time for the three individuals was suggested by several Public Works Department personnel. Since separate wage figures for each of the three individuals were not available, however, the procedure described in the text was used instead.

[f]The dispatcher is nominally assigned to street maintenance and his salary is included in the figures for that department; but, in fact, he dispatches vehicles for all of the operating services. It was assumed that his time could be apportioned according to the number of radios used by each operating service. On this basis 12/80 of the dispatcher's salary and fringes, or $1.674, was attributed to refuse collection.

[g]The five individuals are the Director of Public Works, the Assistant to the Director, the Records and Services Supervisor, one secretary, and one administrative clerk. Their total wages for 1972-73 came to approximately $78,404.

and that the time devoted to refuse collection by those five individuals could be calculated as above (i.e., 71% times 21% = 14.9% of their time). Calculating on this alternative basis yields a total figure of $14,612 for wages and fringes for departmental administration.

Administrative Overhead. The city's accounts record an (actual) expenditure of $7,149 for Public Works Department administration beyond the amounts distributed as wages. However, most administrative cost items other than direct wages are not accounted for in departmental or subdepartmental budgets. This is true for such items as office space, office equipment and supplies, utilities, telephone, insurance, data processing, and so on. Nor is the time devoted by city departments, such as the Controller's Office, the Health Department, and the legal staff, to matters related to refuse collection allocated to that service, or to the Public Works Department as a whole. It was assumed that these overhead items—including indirect labor, such as the legal staff and Controller's staff, as well as office space, office supplies, etc.—came to 33 percent. This figure was based on data for similar expenditure items obtained from four private refuse collection firms (which necessarily must reflect most such costs in their accounts).[h] A minimum estimate might be 10 percent of wages and fringes;[i] a maximum would probably be 66 percent.[j] When applied to the two different estimates for the cost of departmental administrative wages and fringes, these different percentages yield a range of estimates for departmental overhead costs between $1,461 and $17,755.

The best estimate including wages, fringes and overhead seems to

[h]The four companies are subsidiaries of Sanitas Waste Control Corporation; the parent company generously provided the financial data. The data available for these companies, however, was unclear or incomplete in various areas. Furthermore, the companies—which are located in the western United States—may not be entirely comparable to Hartford's operation in significant respects. Finally, the ratios between figures varied from one company to the next (although they were all well within the 10 to 66 percent range suggested here). Consequently, the estimate of 33 percent derived from this data is necessarily crude.

[i]If an individual earning $15,000 per year occupied a 15' × 15' office, the annual rental value of the office alone, at $5 per square foot (probably a relatively low estimate for downtown Hartford), would be $1,125, or 7.5 percent of his salary.

[j]Yale University presently estimates 66 percent of salaries as the cost of overhead on research contracts awarded to faculty.

be $28,869—based on the lower of the two figures for departmental wages and fringes, and using the 33 percent markup estimate.

Capital Costs of Vehicles

The city of Hartford maintains a substantial fleet of vehicles for use in its refuse collection service. This fleet includes; (1) twenty-four "packers"—trucks fitted with 20-cubic-yard bodies for collection and compaction of refuse; (2) four "bulky trucks"—open trucks used to pick up large items such as discarded furniture and refrigerators; (3) one truck equipped to collect refuse from sidewalk trash receptables; (4) six panel trucks for use by route foremen; and (5) two sedans for use by supervisors.[k]

The city makes no effort to amortize the cost of these vehicles over their useful life, nor does it include in any of its budget figures the opportunity cost (imputed interest cost) of the capital funds it has tied up in the vehicles. The following sections describe the methods we used to arrive at the depreciation figures (amortization of purchase price) and imputed interest cost for the city's refuse collection vehicles.

Depreciation. Annual depreciation for each category of vehicles was calculated as follows:

Purchase Price.—The year of purchase and the price paid by the city for each vehicle was obtained from the city's records. These price figures were then inflated to 1972 dollars by means of the wholesale price index computed by the U.S. Department of Labor and reported in the *Survey of Current Business* for "motor vehicles and equipment." The resulting figure, $434,128, represents crudely what it would have cost the city to replace its fleet with new vehicles of a similar type in 1972.

Depreciated Value.—The depreciated value of the city's vehicles as of the beginning of the 1972-73 fiscal year ($305,374) was calculated by taking the purchase price of each vehicle, in 1972 dollars, and reducing it by 11 percent for each year of the vehicle's age.

[k]During the 1972-73 fiscal year, the city had in its possession—unused—several trucks that were formerly used for collection from dumpsters, an operation that was discontinued before the beginning of the 1972-73 year. The cost of these vehicles was omitted from the calculations presented here. So long as these vehicles were not sold or converted to another use, however, they represented a cost properly attributed to the city's refuse collection service and could reasonably have been included as part of the city's total cost of refuse collection.

Depreciation for 1972-73.—The depreciation cost for 1972-73 ($47,754) was computed as 11 percent of the purchase price (in 1972 dollars).[l] This represents the amount that the city must spend each year, on the average (in 1972 dollars), for replacement of vehicles if the city's fleet is to remain in its condition as of 1972.[m]

Depreciation Rate.—From records maintained by the Public Works Department, it was established that the last nine vehicles traded in by the city had depreciated, on the average, 11 percent per year.[n] It was assumed that the 11 percent rate could be applied as well to the other trucks and the sedans.[o]

Interest Cost. Hartford purchases its vehicles for cash, and consequently pays no actual interest on the purchase price of its vehicles. Nevertheless, there is an opportunity cost to such an investment; the funds involved could have been invested elsewhere, by the city or the taxpayers, to bring in an annual return.

This interest cost was computed by multiplying the (depreciated) value of the city's investment in its vehicles, in 1972 dollars, by an imputed interest rate. The rate chosen was 8 percent, which may be somewhat higher than the rate the city would have had to pay if it obtained the funds through the sale of tax-free municipal bonds, but lower than the return that the city or taxpayers might have obtained

[l]The purchase price of one vehicle that had a depreciated value of zero under the 11 percent rate was excluded for purposes of this calculation.

[m]Since the city has twenty-four packers, and keeps them an average of eight years, it should be expected to purchase three new packers each year to maintain its fleet — which, in fact, has been roughly the practice in recent years. The disaggregated figures underlying these calculations suggest that it should have cost the city $42,465, plus trade-in, to purchase the required three packers in the 1972-73 fiscal year. In fact, during that fiscal year the city traded in four eight-year-old packers on three new ones, paying a total of $51,392 plus the trade-in. This figure is probably higher in large part because the three trucks purchased were of higher quality than the other trucks in the city's fleet. (For example, they had a higher compaction ratio, improved transmission and power take-off, and were more durable). Thus, the higher cost of these vehicles largely represents the cost of improving, rather than simply replacing, the city's fleet. This improvement, it might be noted, was not realized during the 1972-73 fiscal year, since the new vehicles were not delivered before the end of that year. Consequently, none of their increased cost seems properly chargeable in 1972-73.

[n]The last nine packers the city disposed of were traded in on purchases of new packers after they had been owned by the city for eight years. The trade-in the city was given was equal, on the average, to 9.5 percent of the purchase price, figured in constant (1972) dollars.

[o]Some support for this assumption—which might seem to attribute too much longevity to small vehicles—can be derived from the fact that one of the two sedans was seven years old at the time the data were collected.

by investing the funds elsewhere. The city's imputed interest cost on its investment in vehicles at 8 percent is $24,431.

Range of Error. What are reasonable upper and lower bounds for the depreciation and interest costs? If the depreciation rate for the city's vehicles is assumed to be not 11 percent, but instead 20 percent (which seems to be the maximum plausible estimate), then the depreciation cost for 1972-73 increased from $47,754 to $73,930.[p] However, because the depreciated value of the vehicles as of 1972 drops from $305,374 to $224,378 under the 20 percent rate of depreciation, the interest cost at 8 percent drops from $24,431 to $17,950. Thus, the shift from an 11 percent rate of depreciation to a 20 percent rate increases the total capital cost for vehicles only from $72,186 to $91,880.

Assuming a depreciation rate of 20 percent per year and an interest rate of 12 percent, the total capital cost for vehicles comes to $100,855. This appears to be the maximum defensible estimate. On the other hand, a depreciation rate of 11 percent per year (which appears to be the minimum defensible estimate) and an interest rate of 4 percent yields a lower bound estimate of $59,970.

Vehicle Maintenance Operation

Direct Labor and Materials. Refuse collection vehicles, like most other vehicles operated by the city of Hartford, are maintained and serviced at a central service and repair shop located at the Public Works Department yard. The city does not allocate to the refuse collection budget the fraction of the shop's expenses that are attributable to refuse collection vehicles. Rather, the city maintains only a separate aggregated set of accounts for the maintenance shop as a whole. For the first six months of 1973, the shop did, however, maintain a record for each vehicle it serviced showing: (1) the cost of parts used in repair of the vehicle; (2) the cost of mechanics' labor used in repair of the vehicle (calculated by multiplying each mechanic's hourly wage by the amount of time he spent working on the vehicle); and (3) the occasions on which the vehicle was greased or had an oil or filter change, or had a tire replaced. This information,

[p]The increase in depreciation cost is not quite proportionate to the increase in depreciation rate, since several of the vehicles had been entirely written off by 1972 under the faster rate of depreciation.

together with the expenditure information recorded for the shop as a whole and some estimates by shop personnel, was used as the basis for estimating service costs for vehicles.

These records indicated that 18 percent of the maintenance shop's costs (labor and materials) were incurred servicing refuse collection vehicles.[q] Applying this figure to the total expenditures on shop operations ($992,549) yielded an estimated cost of $178,659 for parts and labor.

This figure does not include the amortized cost of the shop building itself or of most of the equipment in it, which are not accounted for in the city's annual budget. These costs are estimated in the Buildings, Land and Equipment section, below. The figure does include, however, the cost of gas, oil, and grease, the cost of labor for routine services as well as for mechanical work, the cost of parts, and the cost of administrative and clerical labor at the shop level.

Fringe Benefits. The costs of fringe benefits for shop personnel must be added to the $178,659 figure. The actual expenditure for all wages in the shop for 1972-73 was $578,380 (including administrative personnel). Assuming that 18 percent of the labor represented by this figure can be attributed to refuse collection vehicles, and that fringe benefits cost 25 percent of wages, the cost of fringe benefits attributable to refuse collection was $26,027.

Administration. Since the shop is an operating division of the Public Works Department, it is appropriate to include administrative costs at the operating services and departmental levels as well. These were computed as in Section 2 above, using the ratio of the shop budget to the entire Public Works Department budget (less adminis-

[q]The total cost of parts and labor billed to refuse collection vehicles on the service records for the first six months of 1973 was, respectively, $7,335 and $7,809. Doubling these figures yields full-year projections of $14,670 and $15,618. These figures almost certainly underestimate the actual cost of parts and mechanical labor attributable to refuse collection vehicles. The total parts and labor costs recorded for all vehicles maintained at the shop came to only $82,376 — or, when doubled, $164,752 for the year. The total actual expenditure on direct labor for the shop for the year, however, was $506,899. The latter figure includes personnel who perform routine service on vehicles (tire repair, etc.) as well as mechanical labor. It seems unlikely, however, that the cost of routine service personnel accounts for the enormous disparity between these two figures. Rather than using the figures from the service cards directly, then, those figures were used only to the extent of deriving the ratio between the cost of parts and labor recorded on the cards for refuse collection vehicles and the cost of parts and labor recorded on the cards for *all* vehicles maintained at the shop.

tration) as a means of apportioning the cost of departmental administration. The resulting figure was $39,393.

Insurance. All of the vehicles owned and operated by the city are insured under a single policy for which the city paid $274 per vehicle in 1972-73 regardless of the type of vehicle. The cost of insurance for the 37 vehicles used by the refuse collection service was therefore taken to be $10,138.

Range of Error. The total of shop operating expenses, fringe benefits, higher-level administration, and insurance thus comes to $254,217. If 12 percent rather than 18 percent of total shop costs (calculated as above) are attributed to refuse collection, then the cost figure increases to $172,857. On the other hand, taking 25 percent of shop costs yields an estimate of $349,137. These two figures might plausibly be taken as reasonable bounds for the actual cost of maintaining and servicing the city's refuse collection vehicles, although, given the methods used here to estimate costs and the limited data available, it is difficult to give a very meaningful estimate of the range of error involved.

Buildings, Land and Equipment

It was assumed that 20 percent of the amortized annual cost of the physical facilities at the yard — including land, buildings, garage equipment, and office furnishings — could be allocated to refuse collection. This estimate is based on the observation that 18 percent of the work of the maintenance shop appears to be devoted to refuse vehicles (see preceding section), 22 percent of the space in the major storage shed is devoted to refuse collection vehicles, and 21 percent of the effort of the operating services administrative personnel appears to be devoted to refuse collection (see Administration section, above).

Depreciation Costs. The yard facility was constructed in 1964-66 at a cost of $1,592,000, including garage equipment and office furnishings. Using the American Appraisal Company's average construction cost index published in the *Survey of Current Business*, the cost of constructing the building in 1972 would have been $2,645,000. Assuming a 25-year life for the building and equipment then gives an annual depreciation rate, in 1972 dollars, of $105,800; 20 percent of this figure comes to $21,160.

Interest Cost—Building. Since in 1972 the building was roughly seven years old, its depreciated value (at 4 percent depreciation per year) was $1,904,000. At a rate of 8 percent, the interest cost on this investment would be $152,300; 20 percent of this figure comes to $30,464.

Interest Cost—Land. The yard occupies 15.15 acres of land, at an estimated value in 1972 dollars of roughly $20,000 per acre, or $303,000 in total. At an interest rate of 8 percent, this investment cost the city $24,240 in 1972; 20 percent of this figure comes to $4,848.

Insurance. Insurance to cover fire and other property damage for the yard cost $2,630 in 1972-73; 20 percent of this figure comes to $526.

The total of all of these expenses is $56,998. A lower bound estimate of $25,223 is obtained by assuming that only 15 percent of yard costs should be attributed to refuse collection, that the buildings and their contents have a life expectancy of forty years, and that the applicable interest rate is 4 percent; an upper bound estimate of $94,388 is obtained by assuming that 25 percent of yard costs should be attributed to refuse collection, that the buildings and their contents have a life expectancy of twenty years, and that the applicable interest rate is 12 percent.

COST ESTIMATES DERIVED FROM THE SURVEY

The questionnaire was designed on the basis of the experience gained in obtaining cost data in Hartford and North Haven, Connecticut.[r] In particular, we made an effort to ask only for data we thought could be supplied by the municipalities involved without undue effort, yet would provide an adequate basis for estimating most of the major cost items attributable to refuse collection. As noted below, many of the devices used to project cost figures from the Hartford and North Haven data were employed in the questionnaire.

A copy of the questionnaire and the letter of explanation that accompanied it appear in Appendix C. This section includes some

[r]Since it is similar to Hartford, the analysis in North Haven is not summarized here.

supplemental notes on the type of information requested, and the methods employed in constructing cost estimates from the survey data. The section concludes with a summary of the results of the questionnaire, and a comparison of the cost figures estimated from the questionnaire with the cost figure published by the municipalities in their annual budget reports.

Wages and Fringe Benefits.—The actual expenditure for wages and fringe benefits was requested directly for truck crews, foremen and supervisors, and for administrative and clerical personnel.

Other Administration Costs.—No data were requested pertaining to the cost of office space, supplies, equipment, etc., for administrative personnel. Rather, as in the case of Hartford, these costs were estimated at 33 percent of the total wages and fringe benefits paid to administrative and clerical personnel.

Vehicle Depreciation and Interest.—On the basis of the information requested on the questionnaire, the annual depreciation and imputed interest cost for vehicles were computed using the same method described above for Hartford. The expected life resale value for vehicles was always taken at face value for purposes of computing a straight-line depreciation rate.

Vehicle Maintenance and Operation.—The cost of parts and supplies for vehicle repair, wages and fringe benefits for mechanics, repair work contracted out, and insurance for vehicles were all requested directly on the questionnaire. The cost of overhead for maintenance work was estimated at 54 percent of mechanics' wages.[s] Where the town could supply no cost figure for vehicle insurance — which happened in a number of cases — the cost was simply estimated at $300 per vehicle.

Vehicle storage.—Rather than seeking the actual cost of land and buildings for storage garages, which seemed too burdensome a request, the questionnaire asked only for the number of square feet devoted to garaging refuse vehicles. This figure was then multiplied by $2.50 per square foot to give an estimated rental value for the garage space. Although this approach, like the markup factors employed in estimating some administrative and maintenance costs, was rather arbitrary and had the effect of eliminating a possible source of cost variation among towns, more accurate methods did not appear

[s]This estimate is based upon the experience of New York City. For its vehicle maintenance operation, the city has calculated that the ratio of all overhead items (excluding the cost of parts and fringe benefits) to the wages paid to mechanics at 54 percent.

to be worth the considerable effort that would be required to obtain them.

Table B–2 summarizes the results of the cost estimates derived from the questionnaire for the sample of Connecticut towns. Presented are the average proportions of estimated collection costs allocated to labor, administration, vehicle maintenance and operations, as well as the capital costs. Also presented are the cost distributions obtained from the more intensive analysis in Hartford. The comparison indicates that the collection operations, on average, are less labor-intensive than in Hartford (probably because few towns provide roll-out service), and incur slightly greater costs of administration.

Figure B–1 presents a plot of the budgeted figure for refuse collection for 18 towns and the refuse collection cost estimated in this analysis. The diagram indicates that budget figures themselves substantially underestimate the costs of service provision, as these cost estimates have been derived in this analysis. The average underestimate is about 23 percent, but it varies between 0 and 61 percent. Thus it appears that using budgetary cost data would have involved considerable bias. Moreover, a simple mark-up of budget estimates, while better than no adjustment, would have introduced some measurement error.

Taxes Avoided by Municipal Operation

A privately operated refuse collection service is subject to a variety of local, state, and federal taxes, which do not apply to a similar service that is municipally operated. To gain some idea of the magnitude of the tax "savings" thus available from municipal operation, we prepared an estimate of the additional taxes for which the refuse collection service operated by the city of Hartford would have been liable if it had been under private ownership.

Table B–3 indicates the magnitude of the savings through municipal operation of refuse collection in Hartford. As the table indicates, the total tax savings are almost $134,000 a year. With the exception of property taxes, which would otherwise be paid by private firms to the city, the other "savings" are really transfers from the federal and state governments. These transfers amount to approximately $94,500 or 5.6 percent of collection costs. The sources of the estimates for each type of tax reported in Table B–3 are:

Property taxes are levied by the city of Hartford on real estate and vehicles owned by private businesses at a rate (in 1972) of 81.4 mills

Table B-2. Summary of Estimated Cost Components for Municipal Collection: Connecticut Towns and Hartford, 1972–73 (in percentages)

Category	*Hartford*		*Connecticut towns*		
			Mean		*Standard Deviation*
1 Direct labor					
Crew labor	—		68.5%		11.5%
Foremen	—		4.3		3.6
Total		75.5%		72.8%	—
2 Administration					
Wages & salaries	1.4%		2.2		2.0
Overhead	0.3		0.7		0.7
Total		1.7		2.9	—
3 Capital costs					
Depreciation	4.1		6.5		5.0
Interest	3.6		2.2		1.4
Total		7.7		8.7	—
4 Vehicle maintenance & operation (excluding insurance)					
Storage	—		1.9		2.0
Parts & labor	—		4.2		2.7
Fuel	—		6.2		4.8
Overhead	2.4		2.3		1.4
Total		14.6		14.6	
5 Insurance (all)		0.6		1.1	0.9
		100.1%		100.1%	

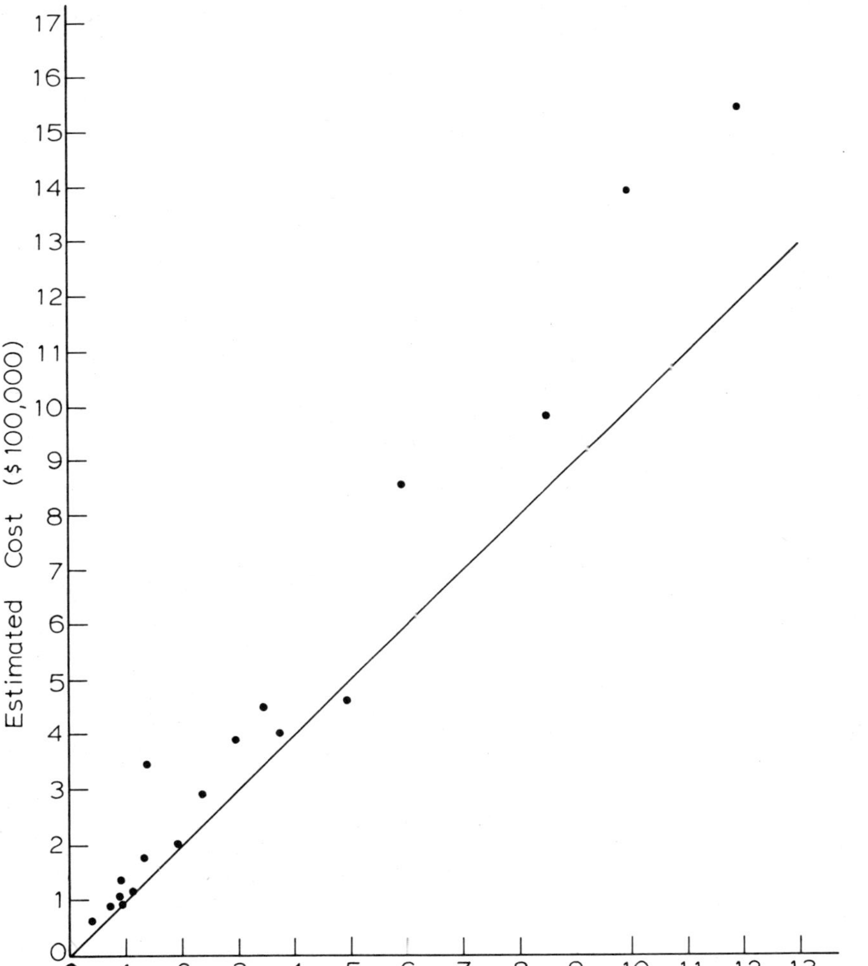

Figure B-1. Municipal Refuse Collection Cost, Estimated Cost Versus Budgeted Cost

per dollar on 65 percent of fair market value. Using the estimates given above for the value of the land, buildings, and vehicles used by the city's refuse collection service, this would come to $39,092 for the 1972-73 fiscal year.

Gasoline taxes are levied on gasoline purchased by private firms at the rate of 10¢ per gallon by the federal government and 4¢ per

Table B-3. Estimated Tax "Savings" Arising through Municipally Provided Refuse Collection, Hartford, Connecticut, 1972

Type of Tax	*Amount*
Property taxes	$ 39,092
Gasoline taxes	15,108
Vehicle registration	6,722
Sales taxes	8,353
Income taxes	64,279
Total	$133,554

gallon by the state of Connecticut. Extrapolating from gasoline consumption records maintained by the city for their packers, and using our own estimates for gasoline consumption by sedans and panel trucks, we estimated that the city's refuse collection service used roughly 107,917 gallons of gasoline in 1972-73. Taxes of 14¢ per gallon on this amount would have totaled $15,108.

Vehicle registration fees must be paid to the state for private commercial vehicles at a rate of 75¢ per 100 pounds for vehicles of greater than 30,000 pounds gross weight, and 50¢ per 100 pounds for smaller vehicles. These fees would total roughly $6,722 for all vehicles operated by Hartford's refuse collection service.

Sales taxes are levied by the state of Connecticut at a rate of 7 percent on most items purchased by private firms. To estimate the amount of sales taxes that would have been collected from a private owner, this rate was applied to: (1) the estimated annual depreciation of the city's vehicles (as an approximation to the amount that was typically spent to acquire new vehicles annually); and (2) half of the maintenance shop operating expenses attributable to refuse vehicles, less gasoline costs (as an approximation to the amount spent on oil, parts, etc.). The resulting amount was $8,353.

Income taxes are levied on private enterprises by the state of Connecticut at a rate of 8 percent of net income, and by the federal government at a rate of 26 percent of the first $25,000 of net income and 48 percent of net income in excess of $25,000. In computing the cost of capital to the city, we have already included an imputed 8 percent opportunity cost of invested capital. This amount might be considered equivalent to the net income, or net return to capital, that would be earned by a private firm in a competitive industry *after* taxes. The problem, then, is to determine the additional before tax return on capital that a private firm would require in order to insure an 8 percent return after state and federal corporate income taxes. Algebraically,

$$T = 0.08I + 0.26\ (25{,}000) + 0.48\ (I - 25{,}000) \qquad \text{(B–1)}$$

where

T is the firm's total federal and state income tax liability
I is the before tax return to capital.

If the after tax rate of return to capital is 8 percent, then

$$I - T = 0.08K \qquad \text{(B–2)}$$

where:

K is total invested capital, appropriately discounted.

Let X be the before tax rate of return to capital so that

$$I = XK \qquad \text{(B–3)}$$

Combining the three equations, and using $738,842 for K (the total depreciated value of land, buildings and vehicles) implies that X = 16.7 percent. Thus an additional 8.7 return on capital, or $64,279, would be required if the service were operated by a private firm subject to income taxes.

Appendix C

Data Collection and Survey Instruments for Intercity Analyses

This appendix describes the procedures and survey instruments used to gather data about refuse collection activities in those political jurisdictions providing municipal collection, in those providing collection by contract, and in those where individual households make arrangements privately with refuse collection firms. A list of jurisdictions providing private, contract, and municipal collection was obtained initially from the Connecticut State Department of Environmental Protection.[a]

Based on these lists of refuse collection organization by jurisdiction, we gathered information about the characteristics of collection activities and their costs by surveys sent to the appropriate official (usually directors of public works departments) in each jurisdiction providing municipal or contract services. For those towns without public collection activities, we surveyed private firms which engage in collection activities and also interviewed town officials.

MUNICIPAL COLLECTION

A survey was sent to mayors or public works directors in twenty-seven cities and towns in Connecticut with municipal collection. The survey instrument attempted to elicit information on the type of service provided within the jurisdiction and the costs of service provision. The disaggregated information on the costs of refuse collection was used to obtain an estimate of the resources actually devoted

[a]We are especially grateful to Charles Kurker of the state Department of Environmental Protection for his assistance in obtaining preliminary information on the organization of collection activities in Connecticut's cities and towns.

to refuse collection in the various jurisdictions. Appendix B documents these calculations. Figure C–1 is a copy of the survey instrument sent to each of twenty-seven jurisdictions.[b]

These survey responses were supplemented by budgetary and wage information obtained by telephone contact in July 1974. This supplemental information was only available for eighteen towns. Telephone interviews, conducted with public works directors or other responsible individuals, were designed to obtain the following additional information: (1) the starting salary for an inexperienced refuse collection laborer on July 1, 1974 and on January 1, 1973; (2) the municipal expenditure (audited figure) for refuse collection activities for the 1972-1973 fiscal year; in particular, officials were asked if the published budget included the costs of:

1. Fringe benefits
2. Vehicle fuel
3. Vehicle maintenance
4. The purchase of new trucks and major equipment
5. Administrative overhead

the officials were also asked in what part of their municipal budgets these costs were reflected; and (3) the municipal expenditure (audited figure) for refuse disposal activities for the 1972-1973 fiscal year.

CONTRACT COLLECTION

Copies of the contracts between local governments and private firms for the collection of residential refuse were solicited from thirty jurisdictions with contract collection.[c] The analysis presented in the

[b]Of the twenty-seven jurisdictions initially contacted by survey in July 1973, one, in fact, did not provide municipal service. In addition, one town refused to cooperate by providing information for our analysis. Extensive telephone contact was required to follow-up unclear or inadequate responses. Data obtained from each town on refuse disposal practices by the state Department of Environmental Protection in August 1972 were compared with our survey results where appropriate, and discrepancies in responses to the few similar questions were investigated by telephone interviews.

[c]Although these contracts were presumably public records, five jurisdictions refused to supply copies of their contracts. In addition, two towns had recently switched to other modes of refuse collection and were not included in the sample. As our analysis was nearing completion, it was discovered that one additional town provided contract collection services sometime during the 1972-1974 period. It is not included in the analysis reported in the text.

Yale University *New Haven, Connecticut 06520*

INSTITUTION FOR SOCIAL AND POLICY STUDIES
111 Prospect Street

YALE/NSF REFUSE COLLECTION SERVICE STUDY

The enclosed questionnaire is designed to gather data for a detailed study of the financing and costs of solid waste collection services within the State of Connecticut. The study is being financed by the National Science Foundation, and is being conducted by the Institution for Social and Policy Studies at Yale University.

The information on the questionnaire should permit us to develop reasonably accurate estimates of the costs incurred by municipalities in providing various types of refuse collection service. Of course, the questionnaire does not cover all of the costs involved, but only the major items. Various overhead expenses, for example, have been left out. However, with the help of extensive and detailed cost figures generously provided to us by the towns of Hartford and North Haven, and with the use of various other industry studies, we believe that we can project reasonably accurate cost figures on the basis of the information requested.

The cost information derived from the study will be used to provide answers to two broad questions: (1) How is the cost of refuse collection related to the type of service provided and the characteristics of the local community? (2) Is it feasible or desirable to finance municipal refuse collection, not out of local taxes, but rather with a service charge for all those who use the service?

Besides being of academic interest, we hope that the results of our study will be of interest to municipalities throughout Connecticut that provide or are considering providing municipal refuse collection service.

Figure C-1. Yale/NSF Municipal Refuse Collection Questionnaire

YALE/NSF MUNICIPAL REFUSE COLLECTION QUESTIONNAIRE

Municipality:____

Name, address, and phone number of the person or persons who completed the questionnaire, or who would be most able to provide further information:

____

____

____

NOTE: A more detailed explanation of the information requested below has been included on separate pages at the end of the questionnaire.

The completed questionnaire should be returned to:

Henry Hansmann
Institution for Social and Policy Studies
Yale University
111 Prospect Street
New Haven, Connecticut 06520

If there are any questions concerning the questionnaire, please feel free to call Henry Hansmann, 436-1087 or 432-4859 (in New Haven).

I. TYPE OF REFUSE COLLECTION SERVICE PROVIDED BY MUNICIPALITY

A. Coverage of Collection

What percentage of the solid waste collected by the municipality each year is:

1. Residential_______________%
2. Commercial_______________%
3. Industrial_______________%
4. Tax Exempt_______________% (churches, private schools, etc.)
5. Government_______________% (municipal offices, public schools, etc.)

B. Residential Breakdown

1. What percentage of the residents of the municipality have their refuse collected by the municipality _______________%
2. If the municipal collection service is not provided to all residents of the city, what criteria are applied to determine who will and who will not receive service? (E.g. city collects from all residents except those living in apartment buildings with more than eight units.)_______________________________________

____

____

C. Type and Frequency of Service

1. How many times per week is refuse collected at each residence? _______________/week.
2. Is refuse picked up:

   At the curb_______________

   From behind the house_______________

   Other (describe)_______________________________________

   ____

   ____

3. Does the municipality collect from dumpsters?______________. If so, what percentage of the refuse collected by the municipality each year is collected from dumpsters?______________%. Describe briefly the number and type of trucks used for dumpster collection:______________

____

____

____

4. Does the municipality collect from containers other than ordinary refuse cans or dumpsters?______________. If so, what percentage of the refuse collected by the municipality each year is collected from such containers?______________%. Describe briefly the containers, and the number and type of trucks used for collection from the containers:____

____

____

____

____

D. Total Annual Collection

1. How many tons of refuse are collected each year by the municipality? ______________tons. (If annual tonnage is unavailable, an estimate of cubic yards collected per year or truckloads collected per year would b helpful:____

____.)

E. Financing

Is the municipality's refuse collection service financed:

1. Through general property tax revenues?______________

2. By a separate charge levied on each household that receives the service?______________(if so, indicate the amount of the annual charge: $______________/year.)

F. Time of Collection

Usual time of day that collection on a typical route begins______________ and ends______________.

II. COST OF REFUSE COLLECTION SERVICE PROVIDED BY MUNICIPALITY

Unless otherwise indicated, all cost figures should represent the total expenditure for the last full fiscal year for which the municipality has actual expenditure data.

A more detailed explanation of the cost information requested below has been included with this questionnaire on separate pages.

A. General

1. The fiscal year used in providing the information below: ____________

2. The name, address, and phone number of the person or persons who completed the questionnaire, or who would be most able to provide further information: ______________________________

   ____

   ____

   ____

3. Do the cost figures given below include the cost of collecting "bulky" trash, such as furniture and refrigerators? __________ If so,

   a. How often is bulky trash collected? ____________________

   b. How many trucks are used for bulky trash collection? __________

   c. How many workers (<u>excluding</u> the driver) work with one truck on bulky trash collection? ____________________

B. Truck Crews

1. Total wages (including overtime and bonuses) paid to drivers and workers on collection crews: $________________________

2. Total value of fringe benefits (such as pension plan, health insurance, disability or life insurance, etc.) accumulated by drivers and workers:

   ________________________ or ________________________%

3. Number of men (<u>excluding</u> the driver) in a typical truck crew: __________

4. Are workers and/or drivers unionized? ________________

C. Cost of Trucks

Year of purchase, purchase price, and capacity (in cubic yards) of each refuse collection truck that the city owned at the end of the last fiscal year:

	Year of Purchase	Cost	Capacity
1.	______	______	______
2.	______	______	______
3.	______	______	______
4.	______	______	______
5.	______	______	______
6.	______	______	______
7.	______	______	______
8.	______	______	______
9.	______	______	______
10.	______	______	______
11.	______	______	______
12.	______	______	______

D. Truck Depreciation

1. The average number of years a truck is kept before being sold or scrapped: ______________________ years.

2. The amount that the municipality can expect to receive for a truck when it is sold or scrapped (in dollars or as a percentage of its purchase price): $______________ or ______________%

E. Other Vehicles

Year of purchase, purchase price, expected life, and expected resale or scrap value (in dollars or as a percent of purchase price) for all other vehicles (such as sedans or small trucks) used by the refuse collection service in addition to those listed in II. C. above:

	Type	Year of Purchase	Purchase Price	Expected Life	Expected Resale or Scrap Value
1.	______	______	______	______	______
2.	______	______	______	______	______
3.	______	______	______	______	______
4.	______	______	______	______	______
5.	______	______	______	______	______

F. Vehicle Storage

1. Does the municipality have a garage or shed in which it houses its refuse collection vehicles? ______________. If not, where are the vehicles housed? ______________________________

____

____

2. If the municipality maintains such a storage garage or shed, the following information would be quite helpful if it can be obtained:

   a. The year in which the building was built: ______________

   b. The original cost of construction for the building: $______________

      (If this figure is not available, an estimate of the amount it would cost to replace the building in 1973 would be sufficient: $______________)

   c. The expected life of the building: ______________

   d. The annual cost of custodial and maintenance work for the building: $______________

   e. The present value of the land on which the building is located: $______________

f. The portion of the garage that is occupied by refuse collection vehicles (as opposed to other city vehicles and equipment): _______%

g. The number of square feet in the building that are devoted to refuse collection vehicles: ______________________________

3. If the municipality rents space in which to house its refuse collection vehicles, what is the annual rental paid? $______________. Does this this figure include maintenance and custodial work for the building? _____ If not, what is the cost of such work? $____________________.

G. <u>Vehicle Maintenance Done By the Municipality</u>

If the municipality does repair and maintenance work on its refuse collection vehicles in its own shop, what is:

1. The total cost of parts for refuse collection vehicles: $____________

2. The total wages (including overtime and bonuses) paid to mechanics for work on refuse collection vehicles: $______________________________

3. The cost of fringe benefits for mechanics for repair work on refuse collection vehicles: $________________ or ________________%

H. <u>Vehicle Maintenance Contracted Out</u>

If the municipality contracts out any or all maintenance work done on refuse collection vehicles, what is the total cost paid for that work: $____________________________

I. <u>Routine Service on Vehicles</u>

If costs for fuel, grease, oil, tires, etc. are not included in G. and H. above, please provide:

1. The total cost of fuel, grease, oil, tires, and other supplies required in the operation of vehicles (aside from those included in G. and H. above): $________________________

2. The total wages (including overtime) paid for labor necessary to refuel, grease, oil, change tires, and work refuse collection vehicles, or provide other service not included in G. and H. above: $______________________________

3. The cost of frings benefits for the labor referred to in the preceding question (I. 2): $________________ or ________________%

J. Administrative Overhead

1. The total annual wages (including overtime) paid to administrative, supervisory, and clerical personnel for work related to refuse collection: $________________________

2. The total annual cost of fringe benefits for administrative, supervisory, and clerical personnel for work related to refuse collection: $________________________ or ________________%

III. REFUSE DISPOSAL COSTS

Does the municipality maintain its own disposal site? __________________

If so, the following information would be appreciated:

A. Dumping Fee

What fee, if any, does the municipality charge for use of the disposal site (e.g., per ton)? $______________________

B. Disposal Costs

1. Does the fee, if any, cover the cost of using the disposal site? __________________

2. If the information is available, what is the cost per ton of operating the disposal site? $______________________

EXPLANATION OF INFORMATION REQUESTED ON QUESTIONNAIRE

We realize that some of the information requested on the questionnaire will not be readily available. Where precise figures are difficult to obtain, an estimate would be appropriate.

The explanations that follow are keyed to the question on the questionnaire by letter and number (e.g. A. 1, C. 3, etc.).

I. TYPE OF REFUSE COLLECTION SERVICE PROVIDED BY MUNICIPALITY

A. 1 - A. 5: A rough estimate may be all that is available here.

B. 1: We would like to know the percentage of residents (or households or living units) rather than the percentage of buildings from which the city collects. Again, we realize that precise figures are probably not available.

II. COST OF REFUSE COLLECTION SERVICE PROVIDED BY MUNICIPALITY

Some of the cost information requested here will undoubtedly be difficult to obtain. We would appreciate whatever is available, even if a number of the blanks cannot be filled in.

The information in section A, B, and C ("General," "Truck Crews" and "Cost of Trucks") is of most importance to us. We would appreciate having this information even if the other figures are not available. Of least importance (though still quite useful, if available) is the information in F. 2 and F. 3 (the detailed figures for "Vehicle Storage").

Some other short-cuts for filling in the form are suggested below with respect to particular questions.

The information requested here concerns only the costs of collection, not disposal (incinerator, landfill, etc.). Some information concerning disposal is, however, requested in Part III of the questionnaire.

A. 3: We would prefer to have cost figures that do not include the cost of collecting bulky trash. Frequently, however, a municipality will handle bulky trash collection and regular refuse collection as an integrated operation, and consequently costs are difficult to separate. If this is the case, simply provide the information requested in A. 3 a., A. 3 b., and A. 3 c.

B. 2: The figure given here should include items such as: 1) the municipality's contribution to a pension plan; 2) the municipality's share of FICA (Social Security) contribution;

3) life insurance premiums; 4) health insurance premiums; 5) contribution or premium for financing workmen's compensation.

The cost of paid vacations and holidays will be accounted for by the figure in B. 1, and need not be considered here.

The figure can be either in the form of an actual dollar cost or a percentage of total wages (that is, a percentage of the figure given in B. 1). For example, if the figure for total wages in B. 1 is $100,000, and if in addition it costs the municipality approximately an additional $25,000 per year for fringe benefits, then either the latter figure or 25% should be entered for B. 2.

E: Vehicles to be included here are, for example, sedans used by supervisors and any trucks or other vehicles used in connection with refuse collection which were not listed under II. C.

F. 1: This information would be appreciated even if the detailed cost figures in F. 2 and F. 3 are not available.

G. 1 - G. 3: Often refuse collection vehicles will be serviced in a municipally-operated garage that also services other municipal vehicles, such as street maintenance equipment and police cruisers. If this is the case, and if separate cost records are not kept for refuse collection vehicles, the figures requested here might be calculated by estimating the percentage of the garage's work that is devoted to refuse collection vehicles, and then multiply the total garage costs for wages, parts, and fringe benefits by that fraction.

If the cost of parts is not available, the figure for wages will be extremely helpful (and vice-versa).

I. 1 - I. 3: The preceding comments with respect to G. 1 - G. 3 are applicable here as well.

J. 1 & J. 2: If some administrative, supervisory, and clerical personnel spend only part of their time on refuse collection, then only the appropriate fraction of their wages and fringe benefits should be included. For example, if the Director of Public Works and his secretary both spend roughly 25% of their time on matters related to refuse collection (as opposed to refuse disposal, street maintenance, etc.), then only 25% of their wages and fringe benefits should be included in the totals given here.

text is based upon information specified in the contract itself, supplemented by telephone query to obtain estimates of the tonnage collected and any other information not available from the contract.

PRIVATE COLLECTION

For those cities and towns providing neither municipal nor contract refuse collection, it was necessary to obtain information on the characteristics of services offered and market prices from individual firms or from a survey of resident households. In April 1974, a survey was mailed to each of the 256 firms in Connecticut that advertised refuse collection services in the classified sections (Yellow Pages) of the telephone books serving any part of the state. A substantial number of these surveys, 22, had to be re-mailed since the firms could not be located at their listed addresses. After as many as three attempts, all but nine of these questionnaires were presumed delivered — at least they were not returned by the U.S. Postal Service.

Figure C–2 is a copy of the questionnaire sent to all private collection firms. Of this initial mailing, we received positive notification that two of the firms were no longer in operation, and that at least three firms in the sample had merged with others which responded to the questionnaire. A total of forty-three firms completed and returned the questionnaire. Of the questionnaires presumably delivered to separate existing firms which advertised their services in the Yellow Pages, the response rate was less than 20 percent.

An attempt to follow-up the mailed questionnaire by telephone suggested several reasons for the low response rate. First, many private collectors expressed the fear that somehow their competitors would learn the rates they charged for service. Other firms judged that the questions asked were too personal or confidential. Although small private firms are understandably reluctant to reveal privileged cost information, firms were only asked to reveal the market prices they charged for providing alternative types of service.

Second, several firms declined to cooperate in the study because they feared that the results of the analysis would shift the refuse collection industry from the private to the public sector.

We supplemented the results of this survey of firms supplying service by a telephone survey of government officials in the towns with private collection. For thirty-seven of the cities and towns

Name of company________________________________

Address of company______________________________ Telephone____________________

Name of person completing questionnaire________________________

Do you collect from <u>private residences</u> in any Connecticut towns?

No____If not, you may ignore the remaining questions. Please return the questionnaire in the enclosed self-addressed stampled envelope. Thank you for your cooperation.

Yes____If so, what percentage of your total business activity is collection from private residences? Please estimate________________

Please list the names of the towns you service at the top of each column and answer the following questions for <u>each town</u>.
(Space is provided on this questionnaire for only four towns. If you have customers in five or more towns please continue on a separate paper).

Name of Town				
A. Do you offer more than one type of service or frequency of collection within these towns				
Yes				
No				
Please list the current charge for each type of service offered.				
Back of house/once a week				
Back of house/twice a week				
Curbside/once a week				
Curbside/twice a week				
Other (describe)________				
____________________				

Figure C–2. Private Residential Refuse Collection Questionnaire

Name of Town				
B. Are there any restrictions on the service: number of cans, restrictions on what will be taken, separation of bottles, etc.				
Yes				
No				
Please describe each restriction and indicate the towns where it applies ______				
______				
______				
______				
C. What is the approximate number of private homes you servé in each town?				
D. Approximately what percent of the rest of your business activity is located in each town?				
E. Does you firm use a town dump or land fill?				
Yes				
No				
For those cases where your own disposal site is not used, please name the dump or land-fill used ______				
______				
______				

F. What was the charge for the same service in 1970?

We realize it may be difficult or even impossible to recall services and charges for 1971 for the various towns. If you are in the situation, please estimate what you can and just leave the rest blank. (Even if you can only estimate the percentage change in costs, it will be helpful).

Name of Town				
Back of house/once a week				
Back of house/twice a week				
Curbside/once a week				
Curbside/twice a week				
Other__________				
__________				
__________				

The completed questionnaire should be returned to:

Ms. Debbie R.S. Stinson
Institution for Social & Policy Studies
Yale University
111 Prospect Street
New Haven, Connecticut 06520

Thank you very much for your assistance.

Questions:

1) Name of official interviewed:

2) a) Do private firms collect all the residential refuse in your town? Explain.
   b) Has there been any "switching" from one type of collection service to another in the last ten years? Explain.

3) a) Can you estimate the number of private firms collecting residential refuse in your town?
   b) Please name the firms. If you cannot name them, please try to name the major firms.

4) a) What types of service are supplied by these private firms? (back-of-house, curbside, once-a-week, twice-a-week collection)
   b) What is the average rate charged a customer for each type of service?

5) What controls does the town maintain over these firms? (e.g., permits, operation fees, submission of annual information, etc.)

6) a) Does the town operate a landfill sight?
   b) Who may use the sight?
   c) How is it controlled?
   d) Does the town charge firms for dumping there?

Figure C–3. Telephone Interviews with Town Officials Concerning Private Collection Practices

which provided neither contract nor municipal collection to all their residents, we telephoned the first selectman's office in July, 1974. Figure C–3 lists the questions asked of first selectmen (or, in some cases, other town officials). These interviews were more successful. Information was obtained on collection costs and the alternative types of service available in twenty-five of the jurisdictions. Most local officials began by quoting their own monthly charges, and volunteered additional information about the collection operations they monitored.

A NOTE ON SPECIAL DISTRICTS

For most of the cities and towns in Connecticut where municipal or contract collection is provided, the coverage of residents is uniform throughout the jurisdiction, regardless of the location of residents' homes. For example, within the city of New Haven, all residents of structures with fewer than five dwelling units are served by municipal collection. For these jurisdictions, gathering the additional information about the characteristics of the areas served — household incomes, residential densities, etc. — for the analysis in the text was straightforward. The areas of the cities and towns were obtained from the *County and City Data Book*, and the rest of the information was available, at the jurisdiction level, from the *U.S. Census of Population and Housing.*

In eleven cities or towns providing municipal or contract collection, however, the service is not in fact provided to all residents. In these towns, special refuse collection districts (usually termed Boros, Fire Districts, or Sewer Districts) had been established to provide collection services to a portion of the residents of the larger jurisdiction.[d] Three of the special districts created to supply refuse collection followed census tract boundaries, so the socioeconomic characteristics of the districts were obtained by aggregating census tract information available through *U.S. Census of Population and Housing: 1970: Census Tracts.* In the eight other cases, the special districts cut across census tract boundaries. In these cases, estimates of the characteristics of the districts were obtained by estimating, for each census tract which overlapped district boundaries, the percentage of the tract's residents served. The estimates were based upon the proportion of the area of the census tract falling within the special collection district.

Wage Adjustments

In order to standardize the cost estimates for differences in wage rates, cost was deflated by a specially constructed wage index. The objective was to estimate what cost would be if each town faced the average wage rate. Thus, the index is defined implicitly by:

$$AC^* = \overline{w}L + rK = I\,(wl + rK) = I \bullet AC \qquad \text{(C-1)}$$

where:

AC^* = wage adjusted (i.e. deflated) average cost
AC = actual average cost
$\overline{w}$ = average wage rate
w = actual wage rate
r = rental rate for capital
L = amount of labor used

[d]For example, within the town of Stonington, residents within the boundaries of the Mystic Fire District, the Pawkatuck Fire District and the Boro of Stonington are served by a contract with a single private firm, whereas all other residents in the town must make their own arrangements with private firms.

K = amount of capital used
I = index

Solving for I and manipulating algebraically, the index can be written as:

$$I = \frac{\overline{w}L + rK}{wL + rK} = 1 + (\frac{\overline{w} - w}{w})\, d_L \qquad \text{(C-2)}$$

where: d_L, labor's share, is given by:

$$d_L = \frac{wL}{wL + rk}$$

Since labor share data are not available for much of the sample, d_L is assumed to be a constant, the average share of labor for the municipal sample. Wage data needed to calculate the index given by equation (C–2) were also unavailable for the private and contract samples. Instead, public data on average hourly earnings of manufacturing workers for 17 regions in Connecticut were used to calculate the index.[1]

For the municipal cost per ton regressions reported at the end of Chapter 2, a better method of standardizing cost for wage differences was used. For the municipal sample, individual wage and labor cost data were available so that labor cost was simply multiplied by the ratio of the average wage to the actual wage. It should be noted, however, that in the regressions in Chapter 3, all wage adjustments, including those for the municipal sample, were done according to the first procedure in order to eliminate potential bias due solely to the deflation procedure.

Tonnage Imputations

As noted in Chapters 2 and 3, tonnage is a better measure of output than the number of dwelling units served. Unfortunately, tonnage data were not available for the private sample. We desired estimates of cost per ton for all types of organization for the analysis of Chapter 3. We estimated cost per ton using both cost per dwelling unit, which was available for all samples, and tonnage per unit, which was estimated using the following equation from the demand analysis in Chapter 4:

$$Q = 0.2794 + .6673Y - 0.1179D + 0.3695FS \qquad \text{(C-3)}$$

where:

Q = tons per unit
Y = median income (thousands of dollars)
D = density (thousands of people per square mile)
FS = family size (number of persons per dwelling unit)

Summary Tables

Table C–1 reports the potential and actual sample sizes. Table C–2 presents the names of the cities and towns in the sample which provide contract or municipal collection service. Table C–3 presents the names of the towns where information about private refuse collection activities was obtained. Figure C–4 shows a map of the state, indicating the location of the cities and towns included in the analysis.

NOTES

1. State of Connecticut, Department of Labor, Employment Security Division, *Connecticut Area Trends in Employment and Unemployment.* Table "Earnings and House by Labor Market Area" (Hartford: State of Connecticut, April 1974).

Table C-1. Summary of Sample Definition.

(1) Type of Service	(2) Number of Initial Mailings	(3) Number Unusable	(4) Number Refused	(5) Number Usable
Municipal	27[a]	1[c]	1	25
Contract	30[a]	2[c]	5	23
Private				
firms	256[b]	14[d]	200[f]	42[g]
town officials (telephone)	39	14[e]	0	25

[a]Source: State of Connecticut, Department of Environmental Protection.

[b]Source: All firms advertising in classified telephone directories in Connecticut, 1974.

[c]Includes towns which had switched from one form of organization to private collection services.

[d]Including 9 firms whose addresses were unknown; 2 firms which were no longer in business; and 3 firms which had merged with others in our sample.

[e]Includes 2 towns discovered to have contract collection and 12 towns where complete information could not be elicited from local officials.

[f]All nonrespondents without positive confirmation of their merger or departure from the market.

[g]Including 25 firms then collecting residential refuse in 45 towns.

Table C-2. Sample Jurisdictions Providing Municipal and Contract Collection.

Municipal Collection

Town (Special District)

1. Bridgeport*
2. Bristol*
3. Derby*
4. East Hartford*
5. East Lyme*
6. Enfield*
7. Groton (Groton City)*
8. Hartford*
9. Meriden (2nd Taxation District)
10. Middletown (Old 1st Sewer District)*
11. Milford*
12. Naugatuck*
13. New Haven
14. New London*
15. North Haven
16. Norwalk (Sewer District)*
17. Norwich (City Consolidated District)*
18. Plymouth
19. Seymour
20. Stamford (Sewer District)*
21. Stratford*
22. Vernon
23. Waterbury
24. Waterford*
25. West Hartford*

Contract Collection

Town (Special District)

1. Ansonia
2. Bloomfield
3. Colchester (Colchester Boro)
4. Farmington
5. Griswold (Jewett City Boro)
6. Meriden (City Schools)
7. New Britain
8. Newington
9. Old Lyme
10. Plainville
11. Rocky Hill
12. Shelton
13. Stonington (Stonington Boro, Pawkatuck Fire District, Mystic Fire District)
14. Torrington
15. Wethersfield
16. Windham (Willimantic)
17. Windsor (Windsor Fire District, Wilson Fire District)
18. Windsor Locks
19. Wolcott

* Additional wage and budget data obtained by telephone interview.

Table C-3. Cities and Towns with Private Collection for which Information was Available.

Avon	Andover	Bethel	Bolton	Bozrah
Bridgewater	Brookfield	Burlington	Canton	Chaplin
Cheshire	Clinton	Cornwall	Cromwell	Darien
East Windsor	Fairfield	Franklin	Glastonbury	Granby
Greenwich	Guilford	Harwinton	Hebron	Killingly
Lebanon	Ledyard	Litchfield	Mansfield	Middlebury
Montville	Morris	New Canaan	New Milford	Middletown
North Branford	Redding	Sharon	Simsbury	Newtown
Norwalk	Southington	Sprague	Thomaston	Somers
Norwich	Watertown	Stamford	Westport	Trumbull
Plymouth	Woodbury	Westbrook		Wilton
South Windsor				
Warren				
Windham				

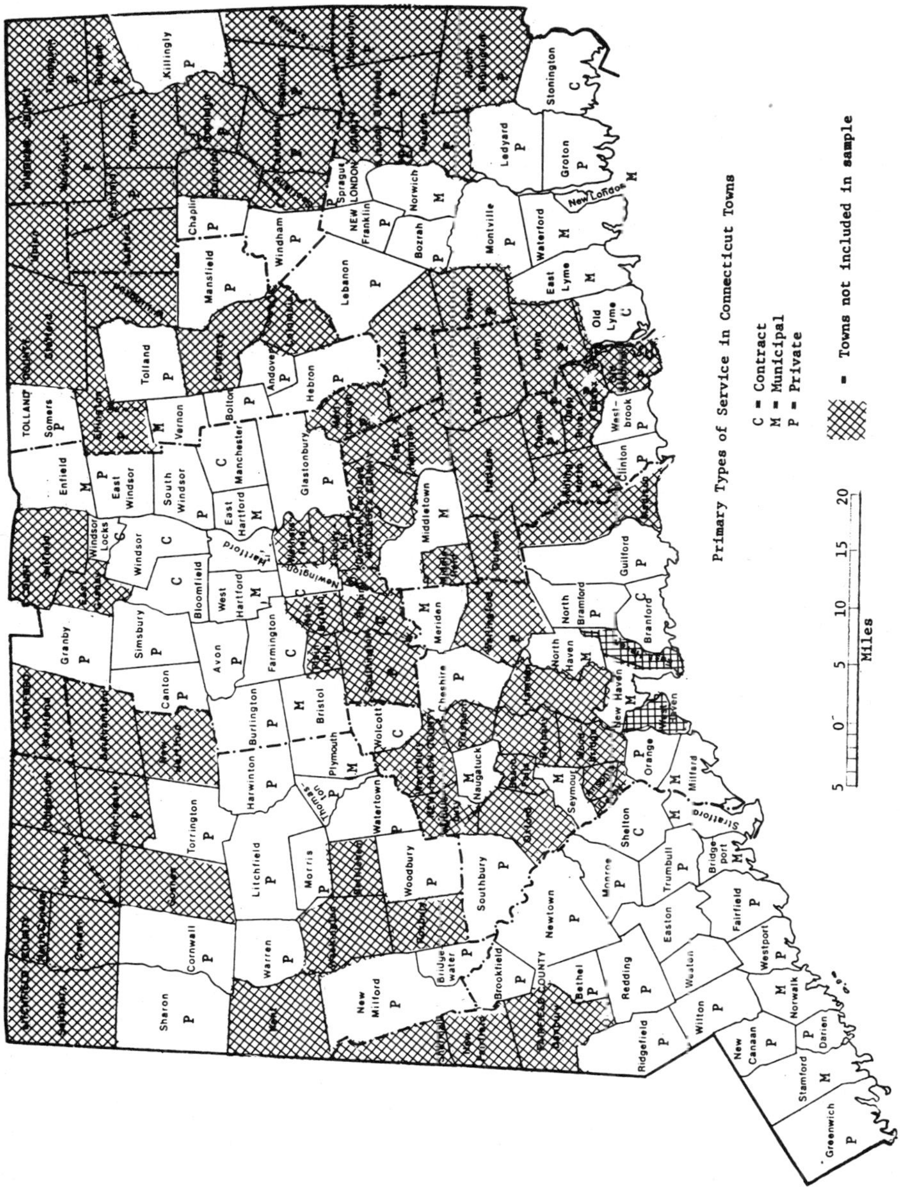

Figure C-4. Map of Primary Types of Refuse Collection Service in Connecticut Towns

Appendix D

Supplementary Tables and Figures

Tables D–1 through D–4 present the means and standard deviations of the variables used in the regression equations in Chapters 2 and 3. Figures D–1 through D–4 present the frequency distributions of the average time variables used in the intracity regressions for Hartford and New Haven in Chapter 2.

Table D–1. Means and Standard Deviations of All Variables for New Haven Sample (number of observations: 2,791 route days)

	Mean	*SD*
Time last load dumped (minutes from midnight)	551.791	56.676
Tonnage dumped	13.364	3.514
Total time per ton	21.807	6.383
Number of runs	2.064	0.524
Round trip haul mileage	3.903	1.318
On interstate highways	2.218	0.892
On city streets	1.685	0.612
Haul mileage per ton	0.620	0.271
On interstate highways per ton	0.349	0.206
On city streets per ton	0.271	0.171
Density (tons per pickup mile)	2.207	0.974
Number of runs per ton	0.159	0.039
Route mileage	6.609	1.169
Snow on ground at 7:00 a.m. (inches)	0.163	0.612
Rain (1 if precipitation, 0 otherwise)	0.256	0.735

Table D-2. Means and Standard Deviations of All Variables for Hartford Sample (number of observations: 519 route days)

Variable	*Mean*	*Standard Deviation*
Total time spent collecting (minutes)[a]	2066.80	253.82
Pickup time (minutes)	1610.60	242.40
Haul time (minutes)	410.87	89.51
Tonnage	78.09	13.44
Total time per ton	27.00	4.28
Pickup time per ton	21.05	3.89
Haul time per run	34.47	5.21
Crewsize per route	24.68	1.91
Number of runs	14.40	1.78
Round-trip haul mileage	8.46	2.39
Round-trip haul mileage on interstate highways	4.42	2.36
Round-trip haul mileage on city streets	3.97	1.48
Pickup mileage	12.55	2.95
Rain (inches)	0.13	0.36
Snow on ground at 7:00 a.m. (inches)	0.22	1.06
Density (tons per pickup mile)	6.67	2.24
Pickup mileage per ton	0.17	0.05
Total haul mileage per ton[b]	1.31	0.43
Interstate highway haul mileage per ton[b]	0.69	0.41
City street haul mileage per ton[b]	0.61	0.22
Number of runs per ton[b]	0.15	0.02

[a]Haul time plus pickup time does not necessarily equal total time because of imputations for missing data (see Appendix A).

[b]The last run for each truck is one-way only.

Table D-3. Means and Standard Deviations of the Variables for Municipal, Contract, and Private Samples[a]

	Municipal	*Contract*	*Private*
Cost per ton	17.22	13.41	
	(6.27)	(7.23)	
Cost per unit	23.28	19.75	41.85
	(7.89)	(6.67)	(13.47)
Service	0.24	0.17	0.55
	(0.40)	(0.37)	(0.50)
Frequency	67.57	62.90	66.24
	(22.74)	(19.34)	(23.33)
Unit density	14.57	10.64	2.35
	(13.76)	(10.30)	(1.71)
Tonnage density	25.30	13.37	
	(26.01)	(12.47)	
Number of observations	23	21	84

[a]For each variable, the mean is listed on the first line; standard deviations are in parentheses below means. For the contract cost per ton regressions, data limitations reduced the sample size to 15 so that means and standard deviations are different from those reported here.

Table D-4. Means and Standard Deviations for Variables in the Pooled Intercity Regressions.[a]

	Once-a-Week		*Twice-a-Week*	
	Back-of-House	*Curbside*	*Back-of-House*	*Curbside*
Cost per ton	22.87 (6.23)[b]	14.96 (5.47)	30.54 (5.35)	18.67 (5.45)
Reciprocal of imputed tonnage density	0.271 (0.200)	0.273 (0.291)	0.418 (0.30)	0.278 (0.300)
Private organization	0.78	0.56	0.85	0.40
Contract organization	0.08	0.23	0.05	0.33
Number of observations	37	57	20	15

[a]These regressions are reported in Chapter 3.

[b]Standard deviations are in parentheses

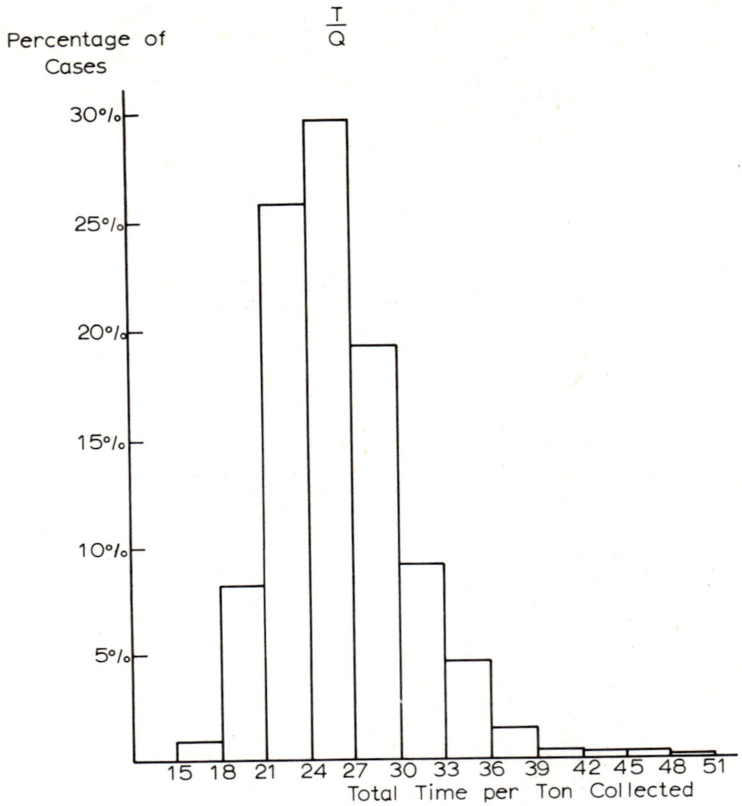

No. of Cases	6	53	169	194	126	59	30	9	3	2	2	1
% of Cases	1.0	8.2	25.8	29.6	19	39.1	4.6	1.5	0.5	0.4	0.4	0.2

Figure D-1. Frequency Distribution of Total Time Spent per Ton Collected for Route-Days in Hartford

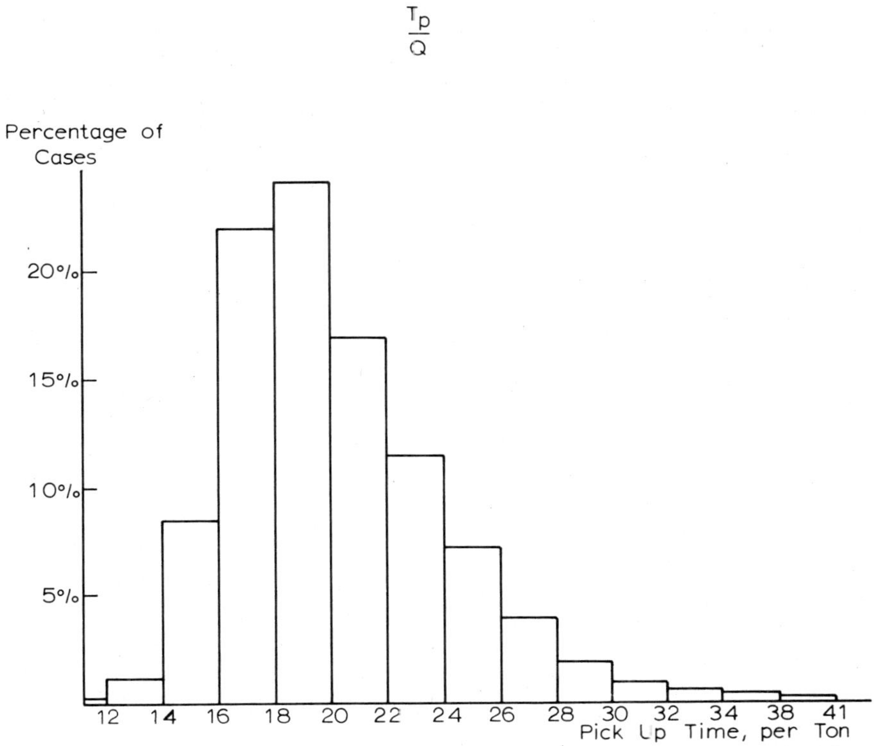

No. of Cases 2	7	55	142	156	110	75	46	26	13	6	4	3	3
% of Cases .4	1.1	8.5	22	24.1	17	11.5	7.1	4.0	2	1	0.6	0.5	0.5

Figure D-2. Frequency Distribution of Pickup Time per Ton, for Hartford

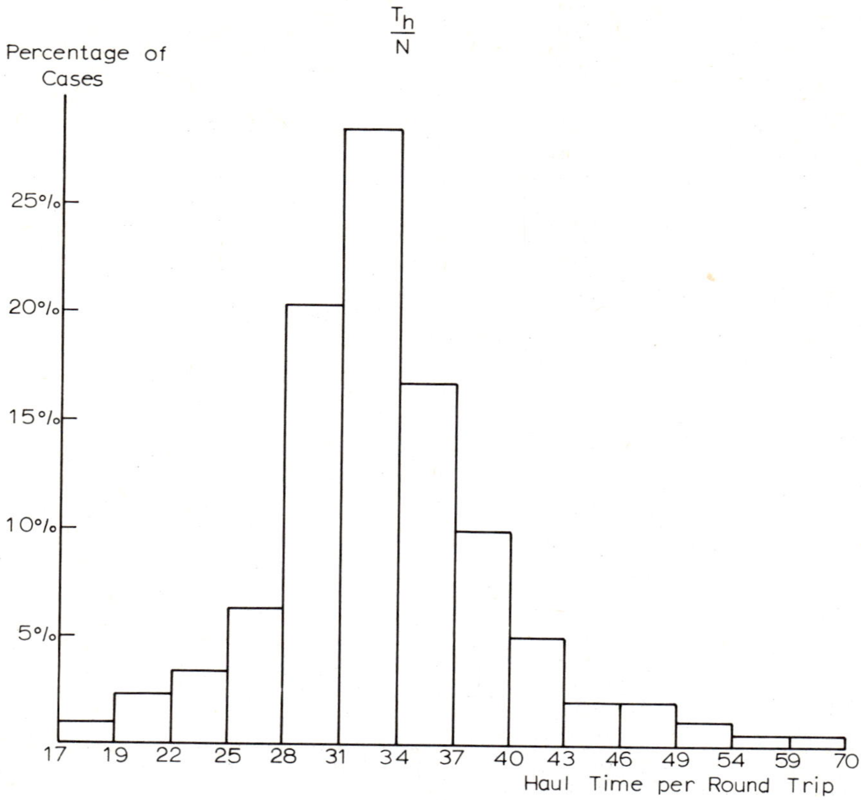

No. of Cases: 6	16	23	41	132	187	108	65	33	12	12	7	3	3
% of Cases: 1	2.4	3.5	6.4	20.4	28.9	16.7	10	5.1	1.9	1.9	1.1	0.5	0.5

Figure D-3. Frequency Distribution of Haul Time per Round Trip, for Hartford

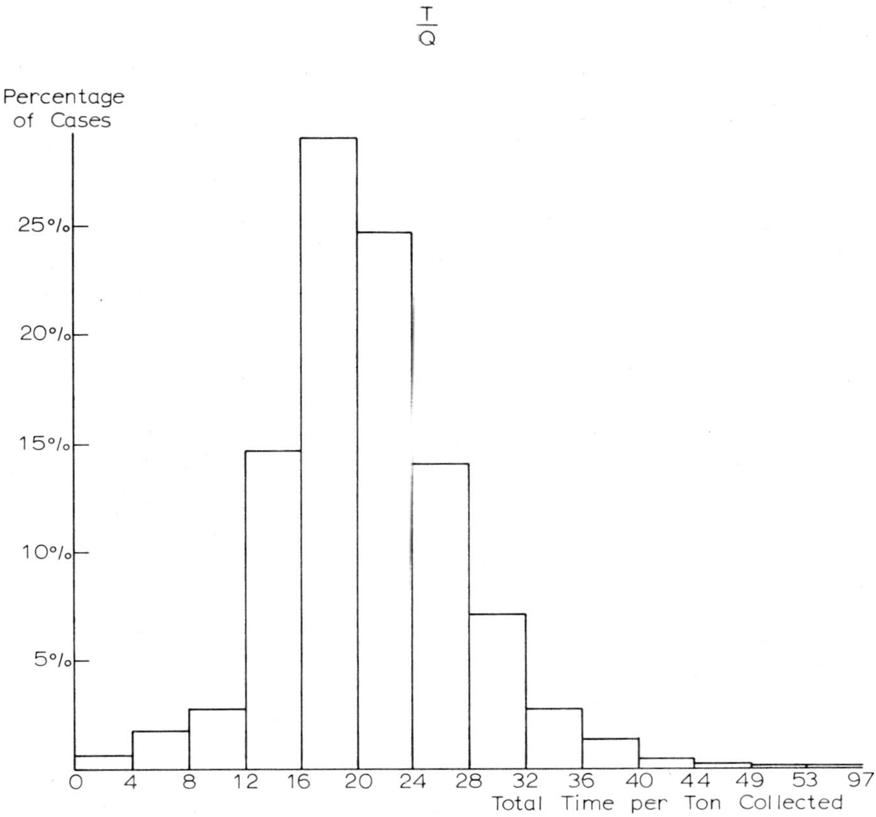

No. of Cases: 17	54	82	42	842	713	409	204	83	41	14	6	4	3
% of Cases: 0.6	1.9	2.8	14.6	29.1	24.7	14.1	7.1	2.9	1.4	0.5	0.2	0.1	0.1

Figure D-4. Frequency Distribution of Total Time per Ton, for New Haven

Bibliography

Ad Hoc Group for Office of Science and Technology. *Solid Waste Management: A Comprehensive Assessment of Solid Waste Problems, Practices, and Needs.* Washington, D.C.: Office of the President, 1969.

American Public Works Association. *Municipal Refuse Disposal.* Chicago: Public Administration Services, 1961.

American Public Works Association. *Refuse Collection Practices.* Third Edition. Chicago: Public Administration Services, 1966.

American Public Works Association. Institute for Solid Wastes. *Municipal Refuse Disposal.* Chicago: Public Administration Services, 1970.

Applied Management Sciences, Inc. *The Private Sector in Solid Waste Management: A Profile of its Resources and Contribution to Collection and Disposal.* U.S. Environmental Protection Agency Report (SW-51d.1). Washington, D.C.: U.S. Government Printing Office, 1973.

Black, Ralph J., et al. "The National Solid Waste Survey: An Interim Report." Paper prepared for the Annual Meeting of the Institute for Solid Waste of the American Public Works Association, October 24, 1968.

Blair, L. H. and Schwartz, A. I., *How Clean is Our City?* Washington, D.C: The Urban Institute, 1972.

Blumenthal, Ralph. "Garbage Has Been a Dirty Racket for a Long Time," *New York Times*, January 5, 1975, Section 4, Page 6, Column 4.

Clark, Robert M.; Gupenhoff, Betty L.; Garland, George A.; and Klee, Albert J. "Cost of Residential Solid Waste Collection," *Journal of the Sanitary Engineering Division*, Proceedings of the American Society of Civil Engineers (October 1971).

Davidson, George R., Jr. *A Study of Residential Solid Waste Generated in Low-Income Areas.* Report SW83ts. Cincinnati: U.S. Environmental Protection Agency, 1972.

Gueron, Judith M. "The Economics of Solid Waste Handling and Government Intervention," in *Public Prices for Public Products*, ed. Selma Mushkin. Washington, D.C.: The Urban Institute, 1972.

Hirsch, Werner Z. "Cost Functions of an Urban Government Service: Refuse Collection," *Review of Economics and Statistics* 47 (February 1965): 85-93.

——. *A Statistical Sketch of State and Local Government Activities.* Berkeley and Los Angeles: University of California Press, 1967.

——. "The Supply of Urban Public Services," in *Issues in Urban Economics*, ed. Harvey S. Perloff and Lowden Wingo. Baltimore: Johns Hopkins Press, 1968.

Houthakker, H. S. and Taylor, Lester D. *Consumer Demand in the United States.* Second Edition. Cambridge: Harvard University Press, 1970.

Kaiser, E. R. "Chemical Analysis of Refuse Components," in *Proceedings of the 1966 National Incinerator Conference.* American Society of Mechanical Engineers, 1966.

Kitchen, H. M. *A Statistical Estimation of an Operation Cost Function for Municipal Refuse Collection.* Ottawa: Trent University, 1973.

Moak, Lennox L. *Refuse Collection and Disposal Service Charges.* Chicago: Municipal Finance Officers' Association, 1961.

Morgan, James N., et al. *Income and Welfare in the United States.* New York, N.Y.: McGraw-Hill, 1962.

Muhich, Anton J. "Sample Representativeness and Community Data," in *The National Solid Waste Survey: An Interim Report*, by Ralph J. Black, Anton J. Muhich, Albert J. Klee, H. Lanier Hickman, Jr., and Richard D. Vaughan. Cincinnati: U.S. Department of Health, Education and Welfare, 1968.

Muhich, Anton J.; Klee, Albert J.; and Britton, Paul W. *Preliminary Data Analysis. 1968 National Survey of Community Solid Waste Practices.* Public Health Service Publication No. 1867. Cincinnati: U.S. Department of Health, Education and Welfare, 1968.

Mushkin, Selma J., ed. *Public Prices for Public Products.* Washington, D.C.: The Urban Institute, 1972.

National Academy of Sciences—National Research Council. *Waste Management and Control. Report to the Federal Council for Science and Technology.* Publication 1400. Washington, D.C.: 1966.

Netzer, Dick. *Economics of the Property Tax.* Washington, D.C.: The Brookings Institution, 1966.

Pechman, Joseph A., and Okner, Benjamin A. *Who Bears the Tax Burden?* Washington, D.C.: The Brookings Institution, 1974.

Ralph Stone and Company, Inc., Engineers. *A Study of Solid Waste Collection Systems Comparing One Man with Multi-Man Crews.* U.S. Public Health Service Publication No. 1892. Los Angeles: U.S. Department of Health, Education and Welfare, 1969.

Resolution adopted by Common Council, City of Middletown, Ct., February 5, 1973.

Rogus, Casimir A. "Refuse Quantities and Characteristics, *American Public Works Association Proceedings.* National Conference on Solid Waste Research, December 1963.

State of Connecticut. Department of Labor Employment Security Division, *Connecticut Area Trends in Employment and Unemployment.* Hartford: State of Connecticut, April 1974.

Town of Darien. "Notice to Refuse Collectors," Darien, Connecticut, November, 1973.

U.S. Department of Commerce. Bureau of the Census. *1950 Census of Population, Occupation by Industry*. Special Report P-E, No. 10. Washington, D.C.: Government Printing Office, 1953.

——. *1970 Census of Population. Detailed Characteristics*. Washington, D.C.: Government Printing Office, 1972.

——. *1970 Census of Housing. Metropolitan Housing Characteristics*. Washington, D.C.: Government Printing Office, 1972.

——. *1970 Census of Housing: General Housing Characteristics*. Final Report HC(1)-A8, *Connecticut*. Washington, D.C.: Government Printing Office, 1971.

——. *1960 Census of Population: Industrical Characteristics*. Washington, D.C.: Government Printing Office, 1963.

——. *1967 Census of Governments*. Vol. 4, No. 5: *Compendium of Government Finances*. Washington, D.C.: Government Printing Office, 1968.

——. *1967 Census of Governments*. Vol 6: *Compendium of Government Finances*. Washington, D.C.: Government Printing Office, 1968.

U.S. Department of Labor. Bureau of Labor Statistics. *Area Wage Survey, 1972*. Washington, D.C.: U.S. Department of Labor, 1973.

U.S. Department of the Treasury. Internal Revenue Service. *Statistics of Income 1971, Individual Income Tax Returns*. Publication 79. Washington, D.C.: Government Printing Office, 1973.

Young, Dennis. *How Shall We Collect the Garbage?* Washington, D.C.: The Urban Institute, 1973.

Index

About the Authors

Peter Kemper is an Assistant Professor of Economics at Swarthmore College and a consultant to Mathematica Policy Research. His recent work on industry location appeared in *The Journal of Urban Economics*; he is currently involved in the evaluation of Support Work, a nationwide manpower experiment.

John M. Quigley is Associate Professor of Economics at Yale University and Research Associate at the National Bureau of Economic Research. He is the co-author of *Housing Markets and Racial Discrimination* published by NBER in 1975 and has published papers in leading scholarly journals such as *The American Economic Review*, *The Journal of Policital Economy*, and *The Quarterly Journal of Economics*.